STARDUST
Our Cosmic Origins

Stephen Welch

Pen Press

First published in Great Britain by Pen Press

ISBN 978-1-906206-96-3

Printed and bound in the UK
Pen Press is an Imprint of Indepenpress Publishing Ltd
25 Eastern Place
Brighton
BN2 1GJ

A catalogue record of this book is available from
the British Library

Cover design Jacqueline Abromeit

For my parents

Peter, who had many interests and loved life
and
Rene, who taught me about the stars

About the Author

Stephen Welch was born in Bedford, England in 1952. He studied maths at Southampton University, switched to a medical career in Scotland for a number of years and then turned to computing as a Systems Analyst. He currently lives in Kent, near the Channel Tunnel, and works as a Business Analyst in the insurance industry. He has many interests including: poetry, chess, go, badminton, tennis and reading.

Contents

Chapter 1

Introduction

On a poet's lips I slept
Dreaming like a love-adept
In the sound his breathing kept;
Nor seeks nor finds he mortal blisses,
But feeds on the aereal kisses
Of shapes that haunt thought's wildernesses.

Percy Bysshe Shelley (Prometheus Unbound)

Prometheus stole fire (knowledge) from the gods and gave it to man.

Let me set out my stall from the start. This book is about knowledge, new knowledge that scientists have given us over just the past 100 years or so. It attempts to bring together threads from many different disciplines of current scientific endeavour and to weave them into a pictorial tapestry, showing the best answers we have yet to the questions like 'What are we?', 'Where do we come from?' and 'What is the nature of reality behind our perception of the world?' Much of this knowledge is about things we can't actually see or feel directly so we create mental models and new ideas to aid our understanding; such are the *shapes that haunt thought's wildernesses*.

Just a couple of hundred years ago the best answer to the above questions would have been that we were created, presumably by the god of whichever religion was dominant in the culture you were

1

born into. We may have had the suspicion that this answer was not altogether satisfactory but we would not have had an alternative. Now we do. The amazing advances in many fields of science over the last few hundred years demand our attention, especially if, like me, you are fascinated by the fact of our existence. We are tiny bits of the universe that are actually aware of the universe and that can ask questions about the universe.

The stardust of the title comes from an old 1960s song called Woodstock by folk singer Joni Mitchell. The fuller quote is 'We are stardust, billion year old carbon'. This turns out to be literally true! The Big Bang, around 14 billion years ago, produced only hydrogen (75%) and helium (25%) in the universe plus a few other trace elements. These two are the lightest elements possible. Over time, clouds of these gasses began to collapse under their own gravity and the first generation of stars were formed. Many of these first stars were massive, much bigger than our sun. Large stars use up their fuel (hydrogen) quickly and when they have exhausted their fuel, they end their lives in a huge supernova explosion.

The pressures within a supernova are immense and under these extreme conditions, other elements can form, in complex nuclear fusion reactions. All the heavier elements in our universe were formed in this way: iron, carbon, oxygen, nitrogen, copper, gold, uranium… to name just a few. The supernova explosions of those first stars seeded the galaxies with heavier elements and like dust they spread through the clouds of primordial hydrogen. Then, when other stars formed as those dust clouds collapsed, there were enough heavy elements to form a few small rocky, earth-like planets round some of them.

We too are made from these heavy elements. We are mainly carbon based but we also need many other elements to make up our complex bodies. So, the rocky planet we stand on and indeed our own bodies themselves are all created from the debris produced in the massive explosions of those early stars. Therefore, we really are stardust, billion year old carbon!

How do we know this? A good question. Just 200 years ago, this was not known. Science proceeds in steps; one discovery is the base for further discoveries. Before they knew about nuclear reactions, the best guess as to the age of our sun was just a few million years, because it was thought to be like a fire. They knew how big it was and then calculated how long a fire of that size would take to cool. When nuclear energy was discovered it was realised that the sun was actually a large nuclear reactor and a more correct age could be worked out for it – about 4.6 billion years. Also, the steps in nuclear fusion needed to create the heavier elements could then be deduced.

But that was just one link in the chain of logic. Another thing they needed to know was 'What are the stars made of?' Again, good question! We can't just go to them and collect some 'sun stuff'! This is interesting… It was discovered that when different things burn, if we look closely at the spectrum of light given off, we see that some parts of the spectrum are much brighter than others. It turns out that each element's burning spectrum is slightly different from each other's, rather like a fingerprint. So if we look at the light from a distant star (however far away) and break it down into its spectrum, it can be worked out what elements make up that star and in what proportions. All without leaving our back yard!

Some people criticise science, saying that if we analyse things too much we lose something. I disagree. A scientist is also a human being with human emotions. There is nothing to stop you seeing the beauty of nature, maybe recalling lines of a poem about the object in question or even wanting to take up your brushes and start painting, but it can only add to the wonder if you also have an understanding of how it works, how it evolved. For example, take this quote from planetary scientist Carolyn Porco:

> *Being a scientist and staring immensity and eternity in the face every day is about as meaningful and awe-inspiring as it gets.*
>
> **Carolyn Porco**

This book aims to provide some of that bigger picture, to add to the wonder and to stimulate the curiosity to learn more. Next time you are out at night, hopefully cloud free and away from streetlights, look up at the stars and be rightly awed by their majesty. But then understand that we now know what each star is made of and that you yourself are literally stardust, born in a supernova explosion. You will be just bowled over. Go back only 100 years and anyone looking up at the stars before then could not have known these things.

Chapter 2

Of Myths and Memes

The great enemy of the truth is very often not the lie - deliberate, contrived and dishonest - but the myth - persistent, persuasive and unrealistic. Belief in myths allows the comfort of opinion without the discomfort of thought.

John F Kennedy

Before looking at the implications of the current explosion in scientific knowledge over the past few decades, let's consider how we would have viewed things in the past. From the earliest times of human history, people have been curious about their origins. Indeed, it is this very curiosity that is one of the defining characteristics of the human species.

Prior to the invention of writing, the history of a tribe would have been handed down verbally from generation to generation and of course would have been elaborated on each time. We are nothing if not storytellers! Each tribe would have had its own tribal history, with the oldest stories being about the origin of the world and how people came to be in it. The need to explain things runs deep in us.

Our brains are designed to be pattern recognition devices, we look for patterns everywhere and are extremely good at finding them, even unconsciously. *[I will expand on this in later chapters.]* Think of looking at clouds and seeing shapes in them, shapes of horses and heads, dragons and angels, we love playing with patterns. One particular type of pattern we tend to look for is 'cause and effect' and we are pleased when we come up with a reason for an event because we then think we understand it.

Early people, confronted with the power of nature, things like earthquakes, storms, fire, flood, drought and illness etc, would naturally ask why? What causes these things? The destruction and devastation brought about by these events look very like the result of some great anger that has been unleashed on the world and it would be a short step to the idea that an unseen entity of great power had been enraged by something. What could they do to appease these gods? Bringing gifts, making sacrifices, worshiping and asking for forgiveness would all be tried.

These tribal myths have arisen in every part of the world. Creation myths, gods, spirits, evil forces, supernatural powers, life after death, reincarnation, fairies, angels, demons, pixies, werewolves, the list is as great and as varied as human culture is diverse.

In 1955 the writer and historian Robert Graves wrote a fascinating two-volume book on the Greek myths. As well as re-telling the myths themselves he wrote about how they relate to the history of the area. Many of these myths involve different gods arguing or fighting with each other, usually with one of them coming out on top. Graves explains that, before writing was invented, the account of a war between two tribes would have been passed down verbally from generation to generation, changing and being embellished each time. Eventually this may well turn into a story about the main gods of each tribe fighting each other, with the god of the winning tribe being triumphant. This is one way that many of the myths of the world would have originated.

When one people or nation conquers another, they usually try and impose their beliefs on the people they now rule. However, it is very difficult to be completely successful in this and the beliefs of the conquered people usually survive in some respect. After many years, the newly integrated society will probably have a mixture of beliefs, myths and gods to worship. It is interesting to try and trace the history of a people and the wars and migrations they have gone through by looking at their system of beliefs. Indeed historians do use this as a tool to investigate and track the spread of people and ideas across the globe.

Creation Myths

One particularly common type of myth is the creation myth. People's innate curiosity naturally leads them to question where they and the world came from, and every culture has its own story to that effect. Creation myths are found all over the world with some places actually having multiple versions of them, usually where that country is home to more than one religion. Here is a just a small selection of them:

From Babylon in the Middle East: Everything starts with a union between Apsu (a sweet water ocean) and Tiamat (a salt water ocean) which gives rise to a number of gods. The gods start to quarrel and fight with each other and eventually a god called Marduk emerges supreme and the other gods accept him as the king of the gods. He divides into two the body of one of the gods he defeated and one part of that body becomes the sky and the other part becomes the earth. Then he takes the blood of another defeated god, mixes it with earth and creates mankind.

From Japan: This starts with the earth young and not fully formed, like floating oil. Then three gods came into existence in the High Plains of Heaven. These were followed by lesser gods including the primal couple Izanagi and Izanami. These two stood on the floating bridge to heaven and stirred the brine below with a jewelled spear. When they lifted the spear up the drops from it created the first land, an island to which they went down and got married. Izanami then gave birth, first to more islands and then to more gods, including the gods of wind, mountains and fire.

From Greece: First of all Chaos came into being and then came Gaia (the earth), Tartaros (the underworld), Eros (desire) and Erebos (night). Gaia gave birth to Uranos (the sky), the mountains and Pontos (the sea). Then she and Uranos created 12 Titans. The Titans coupled with nymphs and each other and produced other demi-gods including Zeus, Atlas, Prometheus, Hades and others. The myths tell of many arguments and fights between all these gods with Zeus coming out as the ruler and declaring that Mount Olympus would be the home of the gods.

From Europe and the Middle East: God created the world in six days. He created day and night on the first day and heaven on the second. On the third, he separated land from sea and created grass and trees. On the fourth day, he created the sun, moon and stars. On the fifth day he created fish and birds and on the sixth day he created animals and a man called Adam to rule over the fish and birds and animals. He then created Eve from one of Adam's ribs and put them both in a garden called Eden. In this garden there was a special tree called the Tree of Knowledge and he told Adam not to eat the fruit of that tree. Later, a serpent persuaded Eve to eat the fruit of the Tree of Knowledge so that they would become gods. She did so and gave some to Adam as well. God was angry and condemned the serpent to crawl on his stomach forever more. He also threw Adam and Eve out of the garden and condemned all women to suffer pain in childbirth thereafter.

From India a Sanskrit myth: The gods discuss how to obtain the elixir of immortality. Vishnu suggests that Devas and Asuras churn up the ocean and the elixir will emerge along with herbs and jewels. The gods uproot Mount Mandara and put it on the back of a tortoise and then they use the snake Vasuki as a rope and churn up the ocean, spinning Mount Mandara around. The ocean turned to butter and out of it rose the sun, the moon and Dhanvantari carrying the elixir. The gods fought over the elixir. Rahu took a drop but Vishnu decapitated him before he could swallow it. Ever since there has been a constant feud between Rahu and the moon which explains the waxing and waning of the moon as it disappears and reappears from Rahu's throat.

From China: The god Pan Gu was the offspring of the vital forces Yin and Yang. He came into being inside a huge primordial egg, growing there for 18,000 years. The egg split and the light part floated up to form the heavens while the heavy part became the earth. Pan Gu held the sky up, growing taller gradually for another 18,000 years, by which time the earth and sky solidified. He was then so tired he lay down and died. His breath became the wind, his voice the thunder, his eyes the sun and moon, his hair the stars, and other parts the mountains, rivers, plants etc.

From the North American Cheyenne: Maheo, 'All Spirit', created the Great Water together with fish and birds. The birds grew tired of flying and took turns diving to look for land. When the coot tried, he returned with a ball of mud and gave it to Maheo. He rolled it in his hands and it grew bigger and bigger until only old Grandmother Turtle could carry it on her back. This became the first land.

From Africa, the Dogon of Mali: The world began with a being called Amma, an egg which was the seed of the cosmos. After vibrating seven times it split and the creator spirit Nommo fell to earth. Then a female twin and four more Nommo pairs followed. Between them they created the sky, earth, day, night, the seasons and human society.

These are just a few of the many, many creation myths from around the world. Of the countries mentioned above the myth described is often not the only creation myth, there are usually a number of them. There are also numerous myths dealing with the origin of humans after the world has been created. *[Many of the above accounts have been drawn from the excellent book **World Mythology** edited by Roy Willis.]*

Introducing Memes

On a different tack, I want here to introduce the concept of **memes**. A meme (pronounced meem) is another name for an aspect of culture that can be copied. In most cases this would be an idea or a belief but it could also be a fashion in clothing or behaviour or it could be a method of making something, e.g. the steps involved in producing a better stone axe. *[If you want to keep things simple just think of the word 'idea' whenever I use the word 'meme'.]*

We need to be clear here of the difference between a myth and a meme, they are not the same. Myths are fictional stories that have been developed over the years and passed on from person to person. They rightly can be called memes as well. But the definition of a meme is wider than that (as can be seen in the paragraph above). Also, memes can be true as well as false, for instance a meme that is true would be 'The sun will rise tomorrow', which obviously isn't a myth. A meme that is false would be 'The moon is made of cheese'. Both of these are examples of a meme. All that is required for it to be a meme is that it is something that can be copied and passed on.

The biologist Richard Dawkins, who holds the position Professor of the Public Understanding of Science at Oxford University, invented the term meme in his excellent 1976 book *The Selfish Gene*. In his terms, a meme is similar to a biological gene in that it is copied from one person to another and it can evolve or change as it spreads. The place in which memes reside is a brain, indeed it is almost as if brains have been designed specifically to hold and pass on memes.

Some memes may be good at surviving and spread widely. Others, that are not good at getting themselves copied, may well die out completely. This of course will only leave memes that are good at getting themselves copied (again very much like biological genes). Memes can be opinions too and it is entirely possible to hold seemingly contradictory opinions at the same time. An example might be 'If I am good to people, they will be good to me' and 'If I am good to people, they will take advantage of me'.

Question. What would it be about a meme that would make it good at surviving? Well it could be a number of things: that it is useful; that it is interesting; that it is cool or fashionable. But there is another way in which it might do well too and that is as part of a memeplex.

Memeplexes

The idea of memes is an interesting one and I will say more about them in a later chapter but here I want to look at just one aspect of them. Individual memes can travel in packs. They can support each other, which in turn aids their own survival. Such a pack in Dawkins' terms would be called a memeplex and the following fascinating example of such a memeplex is based largely on his writings.

Memeplexes often come with a group name, like Marxism or Roman Catholicism for example. The individual memes in the group support each other and this helps them (and the group) to spread and hence survive. (This is rather like a group of biological genes working together to create an organism or body. Working together like this helps the individual genes to survive because the organism is good at surviving.)

Let's look at religion as a memeplex (this could be any religion). It would probably include the following individual memes, among many others (again from Dawkins).

1. The strong conviction that something is true even though there is no evidence for it, indeed even where there is evidence against it. *(This is called faith.)*

2. Faith is a virtue; if you have it, therefore you must be virtuous. *(See how these two memes support each other, in fact the less evidence there is the more virtuous you must be to believe it! Think of the admiration in someone's voice when you hear them say 'His faith is very deep'.)*

3. Mystery is a good thing and it is not good to analyse mysteries. In fact, we should revel in their insolubility.
   (Again, see how this meme supports the other two.)

4. Intolerance towards other faiths and especially to absence of faith.
   (This acts as a barrier to contamination by other memeplexes, again aiding survival of the memeplex.)

5. The urge to convert other people to your religion. Also, you must bring up your children in the same faith.
   (A great way to spread the memeplex.)

6. You are not to question the laws handed down.
   (Again, this helps the spread because it stops the memes from being changed or from evolving so they will be copied accurately.)

7. The belief in an afterlife.
   (People fear death, both for themselves and their loved ones. So, this meme is very powerful because it gives people comfort. It is a great addition to the memeplex and will really help its survival.)

8. If you are good, i.e. follow the laws of the religion, you will be rewarded in some afterlife or other. And if you are bad, you will suffer eternal damnation of some sort.

(Help! I had better be good! Again, see how the memes work together.)

9. God is watching you all the time.

(Help! I had better be good!)

You can see that these memes (along with many others in the memeplex) strongly support each other and make it much more likely that the individual memes will survive and be passed on from person to person because you have to take the whole package together. An individual meme on its own might be questioned and found wanting but when you get a whole lot together, all supporting each other, it becomes very powerful and hard to shake off. This is nicely demonstrated by the fact that very rarely do people choose their faith from the ones available. For the vast majority on this planet a person's religion is strongly dependent on the culture they were born into, it is passed on from generation to generation via their parents.

So what about science, is it a memeplex? Yes (remember memes are not always false, they are just ideas). However, two of the main differences from the religion memeplexes are:

1. It does not depend on which culture you happen to be in. (A Hindu doing an experiment in India will get the same result as an agnostic doing it in France.)

2. Science (unlike religion) actively encourages the continual questioning of its own assumptions. This is one of the main ways science progresses, the questioning of assumptions is actually built into the scientific method.

As Dawkins points out, science involves a standard methodology, which includes testability, evidence, consistency, repeatability, universality and independence of culture. Religion spreads despite a total lack of every one of these virtues. *[I'll return to the scientific method in chapter 11.]*

As religions spread, they become more authoritarian and develop a hierarchical structure. New priests for instance have to toe the party line if they want to progress (or even if they only want to stay within the protection of the church.) Again, these structures help the memeplex to survive and prosper. Organised religion is a very powerful force and is extremely good at perpetuating itself.

Governments and political systems also develop these features. They too do their best to ensure their own continued existence. In fact, you often find that governments and religion are very closely entwined with each other, they are both in the business of ensuring that the population do what they are told and it helps both to survive if they support each other. Very often, in the pomp and circumstance surrounding public events, we see that politics and religion are intricately interwoven to make a compelling spectacle of solemnity and authority.

If we look at the USA, we see a country where the constitution, drawn up by the founding fathers, actively tries to separate religion from the apparatus and laws of the country. But now look at the politicians – they fall over themselves trying to show how religious they are. The memeplex of religion claims for itself the moral high ground, with memes like 'If you are not religious then you must be immoral'! See how powerful that idea is. No wonder the politicians want to be seen as representing it.

When Karl Marx was writing about revolution, for freeing the workers from the subjugation of their rulers, he saw clearly how religion and the state worked together. He called religion 'the opiate of the people', meaning that it kept people happy with their lot. It preached that suffering was natural, even good for the soul, and that there would be reward in the afterlife for those that suffered and didn't complain. Russian communism tried very hard to be an atheist system of government and actively persecuted religion across the country but as we've seen the religion memeplex is very persistent and it continued in secret meetings and in people's own homes. After the collapse of Russian communism, there was a resurgence of the old religions almost immediately.

Other communist countries are officially atheist as well but, out of pragmatism perhaps, they didn't try and wipe out religion as strongly. Take China for instance, officially it is an atheistic, communist country and has been since the civil war in 1949. However, only a small percentage of the population actually is atheist, the rest are a complicated mixture of Animism, Taoism, Confucianism, Buddhism, Christianity and Islam, testifying loudly to its chequered history over the years.

Conclusion to chapter 2

While myths and religions have been around for much of human history, so too has scepticism and the desire to think for one's self. In Greece for instance, over 2,000 years ago, a number of gods were worshiped simultaneously, one for the sea, one for the harvest and so on. A Greek philosopher at the time wrote that 'the people found all the gods all equally believable, while the philosophers found all the gods all equally unbelievable but the magistrates found all the gods all equally useful'!

An important point to make here is that all these myths and religions have one thing in common, and that is their age! They were all created before we had the tools to explore the universe properly. Over the past two centuries, slowly at first but then with an almost exponential growth rate, the body of human knowledge has been increasing in leaps and bounds. What was yesterday's received wisdom can become archaic virtually overnight, making it difficult for our social systems to keep up.

Chapter 3

Deep Space – Deep Time

*Space is big. You just won't believe how vastly,
hugely, mind-bogglingly big it is. I mean, you may
think it's a long way down the road to the chemist's,
but that's just peanuts to space.*

Douglas Adams

As we saw at the end of the previous chapter, it was tools that
made the difference. Before using instruments to investigate the
universe, all we had was our common sense. Everything was
interpreted in human terms and human scales, hence in the myths
the gods always had human or animal form and human motivations
and emotions. We had no conception of the vastness of the universe
or of the microscopic nature of the atomic world. *[Notice the word
microscopic, the name of a tool has become a common everyday
word to describe things we can't see but we now know to be
there.]* Much of what science tells us about the universe is counter
intuitive, it doesn't fit in with our every day common sense. It has
only been through the use of tools that we have been able to probe
further.

Outer Space

One of the first tools to be used to investigate the universe was
mathematics. Take for instance the fact that we live on a round,
ball-shaped planet. Very counter intuitive, we only experience it as

flat. It is commonly thought that Columbus discovered the earth was round when he sailed to the West Indies and didn't fall off the edge! In fact, scholars had known this for many years (even if the general public was unaware). The Pythagoreans of the sixth century BC knew the earth was a sphere as did both Plato and Aristotle in the fourth century BC. There were plenty of clues around, like the fact that the position of the stars change as you move north or south. But around 200 BC, a man named Eratosthenes who was the chief librarian at the great library of Alexandria, actually measured the size of the earth and did it using simple trigonometry!

Eratosthenes heard that, at a place in Egypt called Syene, there was a well at which on midday on the summer solstice the sunlight would shine right down to the bottom. This meant that the sun was directly overhead and a stick in the ground there would cast no shadow. So, on the same day, he put a stick in the ground in Alexandria and at noon he measured the angle from the top of the stick to the end of the shadow. He also knew how far north Alexandria was from Syene. Then it was just trigonometry, you could do it yourself if you wanted to. He came up with a figure for the circumference of the earth of 24,608 miles, remarkably close to the figure we use today of 24,857 miles (around the poles).

Once a tool like maths is invented, people will copy it and its use will spread widely (as befits a meme that is so useful). With the invention of writing, printing, transport and eventually radio, TV and then the Internet, ideas and knowledge are spreading round the world at an increasingly faster rate.

Another counter intuitive fact is that the sun does not go round the earth, it only looks like that because the earth is spinning on its axis. We don't feel this in our bones though, it looks very much like the sun is moving across the sky doesn't it? Next time you are looking at a sunset, however, try to imagine that the sun is not really sinking to the west, what's really happening is that the earth, with you on it, is rolling backwards in space! A major aim of this book is to see past these seemingly common sense ideas and to understand what's actually happening in the world around us.

The next main tool appeared with the invention of the telescope and with this a whole new horizon opened up. Galileo pointed one at the planet Jupiter and saw for the first time in history that another planet also had a moon going round it, in fact he saw four moons going round Jupiter and actually worked out their orbits. He also looked at our moon and was able to make out mountains and valleys on it, before this people had thought the moon was smooth.

With a combination of mathematics and the accurate measurements of the stars and planets people like Copernicus, Kepler, Galileo, Cassini and Newton revolutionised our view of the universe. Instead of everything moving round the earth, which was previously thought of as the centre of things, they showed that the earth went round the sun and that it was just one planet among a number of planets in the solar system which had the sun as its centre. Also that our sun was just one star among many. All of which completely changed our perspective, no longer was the earth the most important body in the universe. At this time, however, they still had no idea of what a galaxy was.

As telescopes got better and more people started using them they found some objects in the sky were still fuzzy blobs, which their telescopes couldn't resolve, they named these items nebulae. These were treated as a nuisance and one man, Charles Messier started to catalogue them in the 1780s in order that people could ignore them and not mis-identify them as comets. Many of these are still known today by their Messier number, like M31 the Andromeda galaxy, our nearest large galactic neighbour and which is in fact a huge spiral galaxy like our own Milky Way.

It wasn't until 1923 that Edwin Hubble, using the 100-inch reflector telescope at Mount Wilson in the USA, was able to calculate the distance to Andromeda by measuring the brightness of a particular type of star known as a Cepheid variable. This proved that Andromeda was outside the Milky Way and so must be a galaxy in its own right. He also showed that other nebulae were galaxies and that the Milky Way was just one galaxy among many.

But Hubble went on to make an even more ground-breaking discovery. In 1929 he showed that all the galaxies were all moving away from each other and that the more distant they were the faster they were moving away. This meant of course that at some time in the past they had all been together. *[I'll expand on this in the chapter on the Big Bang.]* Interestingly, in 1930 Einstein actually visited Hubble at the Mount Wilson observatory. ***Einstein's Theory of Relativity***, published in 1915 had also predicted that the universe was expanding.

Telescopes have continued to improve. We can now look at the universe at frequencies other than just visible light, we have radio telescopes, x-ray telescopes and infrared telescopes. We can measure neutrinos and cosmic rays. We are even building gravity wave detectors. And we have the incredible Hubble space telescope, which has been extremely successful. It, by itself, has revolutionised astronomy. It has looked back over 90% of the way to the Big Bang, it has seen planets around other stars and it has seen stars being born within discs of dust, dust that will also become planets around those new stars.

We are even now building new ground-based telescopes that will be better than the Hubble. It was put in orbit to get above the earth's atmosphere, because that makes ground-based images blurry but now we are building computer controlled adaptive optics into our telescopes to correct for atmospheric distortions. Our tools are getting better and better at an ever increasing rate.

With distances so vast in space, we have to measure things in light years. One light year is the distance it takes light to travel in one year. This is a huge distance because light (the fastest thing in the universe) travels 186,000 miles every second! Light from the sun takes about eight minutes to get to earth so we say it is eight light minutes away. The nearest star to the sun is four light years away! Our galaxy is roughly 100,000 light years across. The nearby Andromeda galaxy is 2,250,000 light years away and that is just a member of our 'local group' of galaxies! Note that light takes time to get to us so if we are looking at a star say ten light years away that light took ten years to get here. That means we are really seeing the star as it was ten years ago when the light started out, so we are actually looking into the past when we look at any stars and galaxies, indeed the more distant the object, the further into the past we are seeing. That's why we say that the Hubble telescope has looked back into the past over 90% of the way to the Big Bang.

So how big is space? Well there are 250 billion stars in our galaxy alone. These numbers are hard to conceptualise so let's try a thought experiment. Suppose that you have a fantastic spacecraft that could visit a star at the flick of a switch. Let's say you visited one star every second, which would mean you would visit 86,400 stars in a day! *[Not giving you much time at each star!]* If you did that every day for the rest of your life (say 100 years) then you would only have visited 1% of the stars in our galaxy, there would still be 99% of our galaxy to explore!

And in the universe there are around 100 billion galaxies!

Think of a sandy beach and of picking up a handful of sand and letting the grains run out between your fingers. How many grains of sand did you pick up? How many on the whole beach? There are more stars in the universe than there are grains of sand on all the beaches on earth! Now think of each grain of sand as a star and remember how far apart the stars are from each other. The next nearest star to our sun is four light years away. At the scale of a grain of sand representing a sun, if you put that grain of sand down on a rock you would have to walk around 30 km to place the next grain of sand – our sun's nearest neighbour!

So, you can see that, although we think of a galaxy as a grouping of stars, it is mostly empty space. Galaxies sometimes collide (when the gravity of a local group of galaxies overcomes the general expansion of space). In fact the Andromeda galaxy is on a collision course with our own Milky Way galaxy; the collision due in about 3,000 million years. When they do collide, however, they will pass straight through each other because they are mostly empty space! (Remember how far apart those grains of sand were.) The nice spiral shapes of our two galaxies will change because the gravity of the two galaxies will affect the orbits of the stars around each other's galactic centres but there will be few actual collisions.

Now we begin to get a glimmer of just how big (and how empty) space is, how far apart stars are and how insignificant earth is. Our planet is not the centre of the universe, not even the centre of our galaxy. It is just a tiny chunk of rock circling an ordinary star on the outer edge of one galaxy among 100 billion galaxies.

Inner Space

The first tool used to investigate the very small was the microscope and this opened up whole new worlds that we had been completely unaware of. Galileo built one of the first working ones and used it to examine, as he says 'many tiny animals with infinite admiration. Mosquitoes are the most horrible among them'. As microscopes became more powerful, it was possible to see things the size of bacteria, a thousand of which would fit inside a millimetre. People could hardly believe what they were seeing. A drop of pond water for example contains a whole ecosystem with amoeba and other tiny animals of different sizes chasing and feeding on each other.

The cells that make up our bodies are quite large compared to bacteria, which are around 100 times smaller than our cells. *[For more on bacteria see the chapter on evolution.]* There is a limit, however, to how small we can see with a microscope using light. This is governed by the wavelength of light, which is too long to resolve atoms or molecules. Atoms are so small that 100 million of them would fit into a centimetre. To investigate the next level down we needed other tools. We now have more powerful microscopes such as the electron microscope, which works at shorter wavelengths than light and the scanning tunnelling electron microscope, which can picture individual atoms. But what about the structure of atoms themselves, what are they made of?

The New Zealand physicist Ernest Rutherford working in Manchester in the early twentieth century used the newly discovered radioactive materials to work out the structure of the atom. He fired alpha particles (produced by radioactive decay) at a thin sheet of gold foil. Most of the alpha particles passed straight through but occasionally some were bounced back. He correctly concluded that the gold atoms were mostly empty space with a very hard core in the middle. In fact, if an atom was as big as a football you would still not be able to see the central nucleus. You would have to make the atom as big as the Albert Hall in London and then the nucleus would be the size of a grain of sand. No wonder the vast majority of alpha particles passed straight through the gold foil without hitting anything, atoms are 99.9999% empty!

Other particles do this too. Neutrinos, tiny neutral particles originating in stars and travelling at the speed of light, pass straight through the whole earth (and us) with only a few actually hitting any atomic nuclei on their way. And this is happening all the time, thousands of neutrinos pass through an every square inch of your body every second. So, you can see that we and the earth we stand on are mostly (99.9999%!) empty space. Very counter intuitive! Things certainly feel solid don't they? *[To see where this feeling of solidity comes from see chapter 5 on forces.]*

As our tools improved a picture of atoms evolved to be a central nucleus of protons and neutrons, surrounded by electrons in orbit around it, rather like miniature solar systems. All atoms are made of these same three particles. The difference from one element to another (e.g. between copper and iron, say) is dependent on how many neutrons and protons made up its nucleus.

This billiard ball model of atoms isn't quite right, however. Again, it is nice to think of solid little particles because that's the kind of thing we are familiar with in our macro-sized world. This is the way science works, we come up with a model which works and then, when we probe deeper, we find we have to refine the model or even replace it with a better one. Science involves the continuous questioning and testing of our basic assumptions.

It was found that you couldn't measure exactly where an electron was and its velocity at the same time. The electrons exist in a sort of probability wave at different levels around the atomic nucleus. Also, the protons and neutrons in the nucleus were found to have an internal structure. They are made of things called quarks. There are a whole family of other particles too. We have met neutrinos (of which there are three types) and then there are muons, positrons, tau leptons and pions to name but a few. Also, there are a family of particles called bosons, an example of which is the photon, which carries the electromagnetic waves like light, radio and x-rays. The latest discoveries are being made with large, expensive tools that collide streams of particles together at very high speed and investigate the debris from the collisions to see what the particles are made of.

Another tool used here is the computer, which is needed to store and analyse all the data generated. As I write this, the largest ever particle accelerator is being built on the border between Switzerland and France. It's called the Large Hadron Collider (LHC) and is due to open in 2008. This will generate 40 million collisions a second, creating enough data each year to make a pile of CDs three times as high as Everest! The organisers, CERN, are creating a worldwide network of around 100,000 top-of-the-range PCs in 40 countries, with a transfer rate of 600 megabytes per second.

And now the latest research suggests that even quarks are not solid objects. The most exciting model at the moment is that all fundamental particles can be represented by tiny loops of strings, vibrating in 10 or 11 dimensions. That is a mathematical, theoretical model and one that will challenge the ingenuity of the practical physicists to find experimental evidence for.

So, if the stuff we are made of is mostly empty space are there any forms of matter that are denser than this? The answer is yes, there are two examples. Firstly, a neutron star. When a star finishes burning all its fuel the outward pressure of radiation from its nuclear furnace is gone and its core collapses under its own gravity. A star the size of our sun will become a white dwarf but if it was a few times bigger than our sun gravity would collapse it down to a neutron star, which is a star made up of only protons and neutrons packed tightly together with their electrons stripped away. A neutron star has about the same mass as our sun but is only about 10 km in diameter; a matchbox full of the stuff would weigh as much as Mount Everest.

A neutron star we can perhaps understand, it is made of atoms with the empty space in them removed. But even odder than that is a black hole. When a star collapses that was many times the mass of the our sun, the gravity is so powerful that it doesn't stop at a neutron star, the collapse just keeps going until the density is, theoretically infinite and it becomes a singularity. Singularities hide inside black holes so we can never see them to check out what they actually look like but we have found black holes. There is a massive one at the centre of our galaxy, millions of times the mass of our sun (and indeed at the centre of most large galaxies) but there are plenty of ordinary (large star size) black holes in our galaxy too.

What makes a black hole black is that light can't get out of it. The gravity of the singularity inside the black hole is so strong that it bends the light back in on itself, nothing can escape from a black hole. So, a black hole itself is actually just the region of space that surrounds a singularity and another name for the boundary of a black hole is the event horizon. We know that a singularity is the densest form of matter possible but what that means is very difficult to conceptualise, all that matter existing as a point source! The strength of all that gravity in one place also bends the very fabric of space and time so it is nothing we have any familiarity with.

With these two examples of matter at extreme densities we see the connection between the very large (outer space) and the very small (inner space). This meeting of theories come under the general term 'cosmology' and it includes the Big Bang theory covered in the next chapter.

Another strange aspect of these basic particle building blocks of our macro world is that we can't decide if they are particles or waves. Depending on how we look at them (i.e. which type of experiment we use) they behave sometimes as solid particles and sometimes as waves. There is a famous test called the two-slit experiment where photons are fired at a piece of card with two vertical slits in it. If a detector is placed behind one of the slits we can see which slit the photon goes through, as if it was a particle. However, if the detector is taken away and a second piece of card is placed behind the card with the slits we get alternate vertical patterns of light and shade on this second card. This is exactly like the interference patterns we get from waves interacting. Even one individual photon will make this pattern, indicating that the single photon has passed through both slits at the same time. This experiment has recently been performed with other particles as well, not just photons.

For us trying to understand what this stuff is we are made of, ideas like this are way outside of our everyday experience. Is matter solid or not at its smallest parts? Then we remember that Einstein showed that mass was equivalent to energy, so perhaps thinking about particles as packets of energy may ease the conceptual gymnastics we are putting ourselves through. That thought may also help when considering black hole singularities. We should remember that all these models are just that, models. Many of them are our best guesses that fit the data our tools provide us with and the best way to get to grips with them is to read widely round the subject. Attack it from different angles rather than just taking one viewpoint. There are many things we don't yet know but this is not a reason to give up and say 'it's magic' or 'god created it'. That would be a cop out, we could have done that at any stage but we have always learnt more each time. In fact, it would be a sad day if there was no more to learn. As scientists say of a problem, 'We are working on it!'

The fact that there are things we don't yet know doesn't invalidate our current understanding. For example, when Einstein came up with relativity that didn't mean Newton's laws of motion suddenly stopped working. Relativity just added to our understanding of the universe and in this case just modified the laws of motion under certain conditions, such as in the presence of a strong gravity field.

Our ingenuity in creating more powerful and more accurate tools at an ever accelerating rate has led to this explosion in knowledge, which continues today with newer machines, computers, satellites, DNA sequencers etc. Before this, all we could do was to make up stories about our world, the universe and where we came from. Now we create theories and we keep testing and testing them (or replacing them if required) to make a more and more accurate picture of the universe we inhabit.

In this section we have probed deep into space in two ways. We have found that outer space is incredibly vast, and so empty that even when two galaxies collide they just about pass straight through each other. Then we found that, looking deep inside matter itself, the atoms we are made of and the earth we stand on are themselves 99.9999% empty space. Now we are going to look deep into time itself.

Deep Time

One of the first attempts to be accurate in measuring the age of the world was by Martin Luther. He used the Bible to calculate ages of all the generations in the Old Testament back to Adam and Eve and he concluded that the world was created about 4000 years BC. This method was expanded on first by archbishop James Ussher and then, in 1654, by John Lightfoot the vice-chancellor of Cambridge University to give the date of the creation as Sunday 26th October 4004 BC, at 9:00 in the morning!

After that scientists started to be more rigorous in their reasoning. As mentioned in the introduction, before nuclear power was known about, scientists calculated the age of the sun by assuming it was a fire and by knowing its size and how long fires take to exhaust their fuel. This came out at a few million years. However, geologists realised that the earth must be much older than this to allow for processes such as erosion to create the landscape we see today. When nuclear reactions were discovered it was realised that this was how the sun worked: it isn't a fire but a nuclear fusion reactor. They were then able to calculate that the sun formed about 4.6 billion years ago. Assuming the earth formed at the same time the geologists were happy.

Four point six billion years just rolls off the tongue doesn't it? But it's an immense stretch of time to try and visualise, it just doesn't fit in with our everyday experience. Dawkins supplies a helpful way to think about it. Think of our history: the Second World War, the First World War, the Victorians and the Industrial Revolution. Then further back: the Normans, the Vikings, the Saxons, the Romans and back to Christ. All that history, around 2,000 years, is only 1% of human history. Our species, Homo sapiens, is about 200,000 years old so all that history from us back to Christ, could have happened 99 more times in the history of our species.

Now, hold your left arm straight out and imagine that from your shoulder to your fingers is the age of the earth. Where on your arm do you think all that 200,000 years of human history starts? Your wrist? Your knuckles? Your finger nails even? Well, take a nail file in your right hand and draw it once, gently across the nail of your index finger. Those few nail filings that float down to the floor represent the whole 200,000 years of human history! Gone in just one stroke! Just think how much time that leaves for the rest of the history of the world without humans in it. We have only been on this planet for a geological microsecond. And crucially from this, see how much time evolution has had to work with.

That is just the age of our sun and its planets. The universe is much older; it is around 13.6 billion years old. This figure did not appear overnight, it has been successively refined over the last 100 years and again our tools played a big part in pinning it down. There were times, not so long ago when different measurements gave different answers. At one point the measurement of the age of the oldest stars we could see looked to be older than the calculated age of the universe but gradually the different theories started to converge. We are still refining these calculations and are planning better and better tools to work with. For more details on this see the next chapter on the Big Bang but for now just soak up the immensity of deep time and how much of it occurred without Homo sapiens.

The other day I took a walk back into the distant past. I live in Folkestone on the south coast of England, not far from Dover, and the stretch of coast between these two towns is dominated by magnificent chalk cliffs 100 to 200 ft high. The chalk is made of the remains of microscopic sea creatures that fell to the seabed when they died. Over millions of years this sediment gradually accumulated and became compacted. This happened between 65 and 135 million years ago when England was near the Equator and dinosaurs roamed across the land. At that time, this particular area was under a warm, shallow tropical sea.

So as I set out on my walk, from the top of these cliffs, down a narrow zigzag path, I was going backwards in time. Each foot of chalk I descended past represented around 360,000 years of sediment that had accumulated under that warm tropical sea. You could see that the chalk was a sedimentary rock because there were clearly visible horizontal lines of different strata running through it, representing slightly different conditions in that ancient sea at different times.

In some strata the chalk was very smooth but in others it was embedded with chalky pebbles of different sizes. When this rougher strata was laid down the water would have been shallower, perhaps even a pebbly beach, while the smoother strata represented deeper water. This clearly showed that the sea level had risen and fallen during the 70 million years this chalk was being deposited and in fact had actually done so many times. This could well represent old periods of global warming, when the ice sheets at the poles melted, raising the sea levels round the world.

Interestingly, looking at the cliffs in more detail, it appeared that the process of the sea level receding, as the world got colder and the ice sheets grew again (locking up the water as ice) may always have been a very gradual affair as the smooth chalk slowly turned into the rougher variety. However, the sea level getting deeper seemed to have been quite sudden in comparison with the boundary from very pebbly to smooth being quite sharp, indicating perhaps a quick melting of the polar ice sheets. Having said that, remember that one inch of chalk represents around 30,000 years; that's like from present day back to when Neanderthals were still around. So, this warming probably wasn't quick in our terms, just in geological terms. Our current period of global warming is really unprecedented in its speed.

At the bottom of the cliffs, looking back up, their size and majesty was overwhelming and the seagulls soared and twisted and turned on the rising air currents as their ancestors had done for years, long before humans had been around to see them. But in addition to the immediate vista was the knowledge that I was looking back millions of years into the past to a time when this area of England was under water in a warm tropical sea near the Equator. Again, just 100 years ago, anyone standing there looking up at the cliffs would not have known that. Then I realised that what I was looking at was only a fraction of the 4.6 billion years of earth history. What other travels had this bit of England made in that time and how many other times had it been under water and then raised up again?

In fact, with some very diligent and painstaking work identifying rock strata, fossils and the magnetic orientation frozen into the rocks when they formed, geologists have been able to piece together and map the slow dance of the continents as they moved round the earth on their tectonic plates, colliding into super-continents, breaking up, subducting, rising into towering mountain ranges and being eroded again, with the sediments forming new rocks with newer fossils imbedded in them. Imperceptible movements that built, over unimaginable timescales, to produce these incredible roving journeys in deep time and which will, in the future, change the continents we are familiar with today into new, unrecognisable configurations, long after we as a race have disappeared. Either we live with knowledge or we live with myths.

For instance, we know that there used to be an ocean as wide as the Atlantic between Scotland and England (these two present day countries were actually part of different continents in the past) and that, as this ocean slowly closed, the Scottish mountains were raised up as part of a mountain range as high as the Himalayas are today. Since then so much time has passed that these mountains have been eroded to their present state, a shadow of their former glory, and a new ocean, the Atlantic, has opened up. Again, we get a glimpse of the meaning of deep time.

Before we started making and using scientific tools we thought that everything in the universe revolved around us and that this nice, solid, comfortable world had been created specially for us. Nothing could be further from the truth. In this chapter we have seen how vast and empty space is and how insignificant our little planet we call home is. Then we learnt that what we see as solid, like the earth beneath our feet, is almost nonexistent when you look at it closely. And finally, that the earth has been getting on very nicely by itself for billions of years before we ever came on the scene and will do so again long after we are gone.

Chapter 4

The Big Bang

There is a theory which states that if ever anybody discovers exactly what the Universe is for and why it is here, it will instantly disappear and be replaced by something even more bizarre and inexplicable. There is another theory which states that this has already happened.

Douglas Adams

We have seen how vast space is and we've also seen that the heavier elements that we are made of were created from lighter ones inside stars when they exploded. But where did these lighter elements and indeed the whole universe come from in the first place? There have been two main answers to this. Firstly, that it has always been there (the so-called 'steady state' theory) and secondly, that it was created at some point in the past (the 'Big Bang' theory). Both answers have their philosophical problems. Our everyday experience leads us to think of cause and effect but a universe that has always been there obviously has no 'cause' to bring it into existence so where did it come from? On the other hand, a universe that has a starting point naturally leads to the question 'What was there before the start?' I'll return to these questions later in this chapter but for now let's examine the evidence.

The evidence gathered over the last few decades overwhelmingly supports the Big Bang theory and indeed this is all the stronger due to the fact that it comes from a number of completely different sources. Also, it arises from a very satisfying mixture of astronomy (the very large), particle physics (the very small) and mathematics (the theoretical). Let's take each of these in turn.

Mathematics

In the beginning of the twentieth century Einstein came up with his theory of relativity. This was a mathematical theory that described the structure of space and time and how these were distorted by gravity and mass. The predictions of how things move through space and time from this theory have been verified by extremely accurate measurements and even today this is our best model of the universe at large scales (at subatomic scales we use the theory of quantum mechanics). A few years after Einstein published his theories of relativity it was discovered that some solutions to his equations showed an expanding universe and that running them backwards gave a point origin to the universe. This was just maths though, did it mean anything in the real world?

Astronomy

This is where Edwin Hubble came in. As we saw in the previous chapter he used a certain type of star, of known brightness, called a Cepheid variable to calculate the distance to the Andromeda galaxy. With the realisation that there were other galaxies in the universe than our own Milky Way, Hubble began systematically searching for them and trying to measure their distances.

As he did this, he noticed something different about the light from these distant galaxies. We saw, in the introduction, that we can break down the light we receive from stars into its component spectrum colours and that we could identify the elements a star is made of because when they burn each element gives off a slightly different wavelength (colour) of light. When Hubble looked at the spectrum of distant galaxies he found that the usual large signature of hydrogen at its specific wavelength was not there. However, there was a similar large spike further towards the red end of the spectrum. This was the same for the other expected elements, they had all been shifted towards the red or even infrared area of the spectrum.

This had indeed been predicted by some theorists. In fact, it was a kind of Doppler effect caused by objects moving away from us at high speed. You can hear this yourself using sound instead of light. When a police car with a siren on comes towards you, it will have a high-pitched note. As it passes you and moves away its sound changes to a lower pitch. The wavelength of the sound is compressed while the car is moving towards you and stretched as the car recedes. The same thing happens to the wavelength of light from a galaxy; if it is moving away from us the wavelength of its light will be stretched and, since colour depends on wavelength, the hydrogen spike moves towards the redder part of the spectrum (just like the pitch of a police siren getting deeper as it recedes). What's more, the faster the galaxy is moving away, the bigger the red shift will be.

In 1929 Hubble measured the red shifts of many of the galaxies that he had already calculated the distances of. He plotted his results on a graph of distance against size of red shift and he found a very nice correlation between the two. All the galaxies were moving away from us (except a few in our local group) but the more distant galaxies were moving away faster than the closer ones. This was proof that the whole universe was expanding.

More recently, we have been able to improve vastly the calculations of distance to far away galaxies. Individual stars like Cepheid variables could only be picked out in relatively nearby galaxies, something brighter was needed to see further. Supernovas were bright enough to see individually in very distant galaxies but they could vary in brightness depending on the size of the star that was exploding. Then it was realised that there were different types of supernova and that Type 1a would always be the same brightness. This is where a white dwarf star is in a binary system with another star and it gradually sucks in material from its companion. As soon as it reaches a critical mass, called the Chandrasekhar limit, it explodes as a supernova Type 1a. It does not matter what mass the white dwarf was to start with, it always explodes at the same point and hence with the same brightness. From there, if you measure its apparent brightness you can calculate its distance. This greatly extended the accuracy of distance measurements to remote galaxies and together with the red shift measurements, further proved that the universe was expanding.

Elements

Cosmology is a very satisfying marriage between the incredibly small (particle physics) and the unimaginably vast (astronomy).

From what we have learnt about how atoms are constructed and how they react under extreme conditions, and by using the mathematical equations relating to these, it is possible to wind back the clock to the beginning of time, to the first few seconds and even microseconds of the universe and to calculate how the universe evolved when it was young (even though we don't know precisely when that was, our best estimate at the moment is that the universe is about 13.7 billion years old).

To start with, the elementary subatomic particles were packed together in a plasma so tightly and under so much energy that actual atoms could not form. As the plasma expanded and cooled these subatomic particles were able to start joining up into atoms. Quite a lot of these were anti-matter atoms and when they came together with normal atoms they annihilated each other in explosions. For some reason (still being debated) there was slightly more matter than anti-matter and this survived to give us the universe we know today.

When this process had settled down, at around 300,000 years, our calculations predict that the atoms created would be mostly hydrogen (75%) and helium (25%) the two lightest possible elements, together with just a trace of slightly heavier elements. When we use our instruments today to measure the current relative abundance of different atoms in the universe we find it is made of almost 75% hydrogen and 25% helium, an incredibly close match to our theoretical calculations based on particle physics! This is another, independent, piece of evidence supporting the Big Bang model.

As we have seen, all the heavier elements, such as those you and I are made of, were subsequently created in the supernova explosions of large stars. That too is verified through our knowledge of particle physics about how atoms combine to create heavier atoms (nuclear fusion) at very high temperatures and pressures.

Background Radiation

In 1963 two physicists, Penzias and Wilson, at Bell Telephone Laboratories in the United States, were given the job of discovering why a radio antenna used to communicate with a satellite had quite a lot of static noise interference on it. They found that the noise was the same whatever direction the antenna was pointed so they concluded that it must either be coming from beyond earth's atmosphere or from the antenna itself. They found that pigeons were nesting in it but even after removing them and cleaning out the antenna the noise was still there.

Then they heard that another physicist had been working on how the temperature of the universe might have changed over time if the universe had started with a big bang. After coming up with a figure for how this temperature would appear now, he was starting to build an antenna to detect it. They gave him a ring and when they compared notes it was realised that their 'interference' was actually the afterglow of the Big Bang.

This came from the time, 300,000 years after the Big Bang, when the universe had cooled enough for atoms to form without photon radiation breaking them up again. This photon radiation was then able to travel freely through the universe. It has since cooled to around just 3° above absolute zero and has been red-shifted into the microwave frequencies of the spectrum. It is now known as the cosmic microwave background radiation. The fact that it has been found is another, independent, piece of evidence that the universe once had a hot beginning in some kind of big bang.

One problem, however, was that the radiation was exactly the same in every direction, implying that the early universe was incredibly smooth and uniform at that time. If that was so, then the early atoms would not have been easily able to start clumping together under gravity to make stars and galaxies. Given how quickly galaxies seemed to form there must have been some slight, initial variations in density and this should show up in the cosmic microwave background.

As our instruments improved and we measured the radiation at finer and finer scales, we still found incredible smoothness in all directions. Some people were starting to think there was something wrong with the theory. The instruments were still being refined to try and detect variations at even smaller scales and one scientist at that time memorably said, 'We haven't yet proved that we don't exist.'

Then, with the data returned from a satellite called COBE in the 1990s, we were finally able to see the tiny variations in the background radiation (about 1 part in 100,000). These variations showed the initial clumpiness in the early universe that were to be the seeds of the galaxies we see today.

Summary

There are a number of things we don't yet know about the Big Bang and there are a number of different theoretical models of how it originated and how the expansion unfolded. One set of these theories includes an episode, near the very start of the Big Bang, known as Inflation. This is where something caused the early universe to go through a short period of expansion that was faster than the speed of light. Also, we don't know whether the universe will keep expanding forever or slow down and contract again to a 'big crunch' under the influence of gravity. Current theory, backed by the most recent measurements, favours the expand forever scenario and even that the expansion rate is increasing.

There was one widely reported anomaly in the 1990s where our theories on how individual stars worked was at odds with the calculated age of the universe. Some really old stars were thought to be older than the universe. As more measurements were made and as they became more accurate, this anomaly soon disappeared. It was partly fuelled by media hype. Ever desperate for a good story, they picked up on the results of early calculations but completely ignored the margins of error that the scientists had provided to go with those calculations.

These things are details though. The evidence for a currently expanding universe and an initial big bang of some kind is overwhelming and is supported, independently, from many different branches of science, as shown above. That brings us back to the philosophical questions such as 'What happened before the Big Bang?' and 'If the universe is expanding, what is it expanding into?' i.e. what is outside its current boundaries? These have to be theoretical or philosophical questions because our instruments have a limitation. Some distant parts of the universe are moving away from us so fast that the light from them will never reach us.

Science answers physical questions, it doesn't answer philosophical ones. However, it can rule out ideas as we discover more. For instance, we now know that the sun does not go round the earth, which rules out a large number of ancient myths. There are some areas that science just can't reach and 'before' the Big Bang or 'outside' the universe are examples of these. Having said that, just because science can't reach there is not a reason to suddenly resort back to myth or magic as our ancestors might have done in the face of things they couldn't understand. Science can give us some guidelines here and indeed it would seem logical that any theorising should invoke scientific principles. Why should the rest of the universe be logical and these areas not be?

Run the expansion of the universe backwards and we get to a point singularity and it is here that our equations and theories stop working. The scientist Stephen Hawking, who has done a lot of theoretical work on point singularities and the black holes that surround them cloaking them from view, provides an interesting way of thinking about these questions. He says that maybe the question of what happened before the Big Bang makes no sense because maybe there was no 'before'. Consistent with our theories is an interpretation that both time and space were themselves created in the Big Bang, so there could be no 'before' and no 'outside'.

He gives the analogy of moving north on the surface of the earth. If you keep going in a straight line, you will pass the North Pole and be heading south again. So, you ask the question 'What is north of the North Pole?' Answer, nothing because you can't get north of the North Pole. The question itself is meaningless. Now apply it to the Big Bang and ask 'What is before the Big Bang?' Answer, nothing because time itself was created in the Big Bang, the question is meaningless. Maybe if you headed backwards in time to the start of the Big Bang and kept going you would find yourself heading forwards in time again. Similarly it makes no sense asking 'What is the universe expanding into?' because space itself (our familiar three dimensions) was created in the Big Bang.

In fact, a more accurate view of the Big Bang is that the galaxies are not really moving away from us, they are not expanding into something, they are actually stationary (except for local movements). It is the space between them that is stretching or expanding. Also, it is this expansion of space itself that creates the red shift by stretching the light waves as they travel through it. The Big Bang was not like a bomb going off, it is not shooting the galaxies out from some central spot, it is space itself that is expanding. Think of a cake in an oven; as it rises the raisins in it are getting further apart from each other but they are not moving through the mixture, it is just that the cake itself is expanding.

These are just some ways of looking at these questions that science can't answer directly. There are other speculations, such as the idea that after the expansion the universe will collapse into a big crunch under the power of gravity and then that it might rebound in another big bang and so on ad-infinitum. Although the latest measurements indicate that gravity is not strong enough to overcome the expansion.

Now it might be considered by some that the Big Bang is the last refuge for a creator of some sort. Indeed creationists will probably pounce on this because they love mysteries, so long as they stay mysterious of course. (Scientists like mysteries, on the other hand, because then there is something for them to solve.) 'Ah ha,' they will proclaim with glee, 'you can't answer that question with your science, therefore my answer (that God created it) must be right!'

The illogic in this is obvious: they are maintaining that because science can't (yet) answer a question this means that their answer must be correct, i.e. if 'a' is false therefore 'b' must be correct, case proved! This of course completely ignores 'c', 'd', 'e', 'f', 'g' etc... There is no proof here, just wishful thinking based on a previously held prejudice. Scientists are more honest, saying, 'We don't know yet but we are working on it.'

There are a number of such (scientific) speculations, many of the ideas being only analogies, which of course should not be stretched too far. But the best way to gain some understanding of these questions is to read widely round the subject, this should provide a broad basis for being able to discern reasonable speculation from wishful thinking, myth or magic.

Chapter 5

Fundamental Forces

We had soared beneath these mountains
Unresting ages ; nor had thunder,
Nor yon volcano's flaming fountains,
Nor any power above or under
Ever made us mute with wonder.

Percy Bysshe Shelley (Prometheus Unbound)

As stated before the main aim of this book is to outline the best answers we have yet to the question 'Where do we come from?' Alongside this, however, is a second theme, the objective of which is to investigate the reality of the world beyond our immediate perceptions. When we see the world, what we are seeing is not 'reality', it is a constructed model of the world within our brains. This model has been created and shaped by evolution to be that which is most useful to us to aid our survival in the world. *[More of this in chapter 10].* For instance, we don't see gravity, only its effects such as things falling. Theoretically, we have inferred its existence and have used our tools to measure it but in the past its effects were just taken for granted, built into our internal model of the world without question. If you tripped you would fall, anything else was just (literally) unthinkable.

So, now that we can measure things with our tools, what other forces exist that we are unaware of? Magnetism is probably the one that initially comes to mind, but that seems to us to be a bit of an oddity, a curiosity, useful perhaps to navigators but not for much else. Little did we know! This is the force that keeps us from sinking through the floor! Indeed without it we wouldn't exist and neither would we exist without the other forces of nature. In fact, there are actually four forces in the universe. They are, in ascending order of strength: gravity, the weak nuclear force, the electromagnetic force and the strong nuclear force. These four forces together control the interactions between individual subatomic particles and the large-scale behaviour of all matter in the universe. I'll look at each in turn.

Gravity

Gravity is by far the weakest of the four forces, although conversely it acts over very great distances (which the nuclear forces don't). It is essentially an attractive force that acts between individual particles but, on a larger scale, it governs the behaviour of solar systems and galaxies. It gives us weight, it stops the air we breathe drifting off into space and it keeps the earth in orbit round the sun. In fact, it caused the sun and all other stars and galaxies to form in the first place as the hydrogen and helium atoms from the Big Bang started to clump together under gravity into larger structures. Without it, we definitely would not be here at all.

Although it is the most familiar of the four forces, it does have some counter intuitive properties. Galileo showed that different sized masses fall to earth at the same rate as each other. Einstein's equations show that gravity is actually a distortion of space due to the mass of an object. We can imagine this if we think in two dimensions instead of three. Think of two-dimensional space as an infinite sheet of rubber. When a mass is placed on it, the sheet sags at that point, the bigger the mass the bigger the depression. If anything comes near one of these depressions it either falls in or its path is bent as it goes past, depending how fast it's going. Something as massive as a black hole distorts space so much that even light, the fastest thing in the universe, gets bent by it. In fact any light inside a black hole can't get out because the gravity is so strong, that's why the hole is black, there is no light coming out.

Electromagnetism

When electricity was first discovered it was thought to be different from anything else we knew about but as this was investigated in more detail it was realised that electricity and magnetism were two sides of the same coin. The Scottish physicist James Clerk Maxwell proved this mathematically in 1864 and his equations are still invaluable to us today. This equivalence is quite familiar to us now in simple practical applications such as an electric door bell where electricity flowing through a coil of wire acts as a magnet or, in reverse, a dynamo bike light where electricity is created (induced) in a coil of wire spinning in a magnetic field.

Like gravity, electromagnetism is a long-range force but because it is both attractive and repulsive (unlike particles attract, like ones repel) matter is essentially electrically neutral at a large scale. It is, though, many times stronger than gravity, the electrostatic repulsion between two protons for instance is a trillion, trillion times greater than the gravitational attraction between them!

Maxwell also showed that light itself is actually a form of electromagnetic radiation and he predicted other types of radiation beyond visible light. When our instruments improved we soon proved him correct as we discovered infrared, ultraviolet, x-rays, radio waves etc. All these are the same as visible light but just with different wavelengths of electromagnetic radiation. In fact, if we take the whole electromagnetic spectrum, visible light is just a small portion of it. The question arises, 'What are we missing in our perception of the world because we are limited to visible light?' Some animals, mainly insects, can see ultraviolet and some flowers that need to attract them do have ultraviolet colours, so our view of 'reality' is already found wanting.

This electromagnetic force holds electrons, which are negatively charged, in their orbits round the positively charged nucleus of an atom so it is this force that actually holds atoms together. Light is emitted from an atom when a charged particle is accelerated near it or when an electron drops down from a higher to a lower energy level in the atom.

As well as holding atoms together it is this force that holds molecules together, indeed the whole of chemistry (how atoms and molecules interact with each other) is a result of this force. It is electromagnetism that holds you together, through your atoms and molecules interacting with each other; it also holds your chair together so you don't fall through it (remember the atoms in both you and the chair are 99.9999% empty). When a hammer strikes a nail the actual protons, neutrons and electrons that make up the atoms of the hammer and nail don't touch each other. Electromagnetism acts as a kind of force field and when two large-scale objects (i.e. hammer and nail) touch, it is their force fields that bang together. And because electromagnetism is many orders of magnitude stronger than gravity, the electrical bonds that hold your chair together and the electrical bonds that hold the floor together stop your chair falling through the floor even though the gravity of the entire earth is pulling on you.

There are some particles that whiz through space and are electrically neutral, neutrinos for example. These will pass straight through you and through the earth beneath you (remember how empty things are that we perceive as solid) without interacting with any atoms at all, because they don't feel the electromagnetic force. In fact millions of such particles do pass through us every second and we never know it.

So the solidness we perceive in the world is actually an illusion to do with force fields but a very useful one. It stops us from hurting ourselves, for instance we try to avoid running into brick walls! If we think about it, however, this should not come as too big a shock, we are all familiar these days with radio waves coming into our houses through brick walls and x-rays that see through our bodies. Electromagnetism is responsible for all of the following: our atoms holding together; chemistry; light; combustion; friction and many other familiar physical phenomena. Again, we would not be here without it.

The electromagnetic force is carried by photons. These can be looked on as either waves or particles. *[For a discussion on this counter intuitive idea, refer back to chapter 3, Deep Space – Deep Time.]*

The Weak Nuclear Force

This force is weaker than the electromagnetic force but still much stronger than gravity. It has the shortest range of the four forces. Think of how small an atom is, then remember that that atom is 99.9999% empty with a tiny nucleus at the centre, the nucleus being made of protons and neutrons; well the range of the weak nuclear force is just one thousandth of the diameter of one of those protons.

Neutrons and protons are made of something even smaller called quarks. (The physicist who named them had to call them something and he picked on this strange name from the James Joyce novel *Ulysses*). Quarks come in different types and the weak nuclear force changes one type or flavour of quark into another. There are a number of different types of reaction but essentially it can cause a neutron to turn into a proton and vice versa.

This is crucial to the structure of the universe in that without this force, the hydrogen isotope deuterium could not form and deuterium is necessary for fusion to take place so that the sun can burn. This force is also needed for the build up of the heavier atoms (of which you and I are made) during supernova explosions.

The Strong Nuclear Force

The strong nuclear force, as you might suppose, is the most power-ful of the four forces. It too has a very short range of action, al-though reaching a thousand times further than the weak force it still only acts within the nucleus of an atom. This is the force that holds an atomic nucleus together. If we remember from the electromag-netic force that like charges repel, we can see that the strong nu-clear force must overcome the electrostatic repulsion between the positively charged protons in the nucleus. In fact, this force acts on the quarks inside the protons and neutrons of the atomic nucleus.

The strong force binds the nucleus so tightly together that when lighter nuclei are fused together (fusion reaction) or heavy nuclei are broken apart (fission reaction) huge amounts of energy get released. This is therefore the source of the vast quantities of energy released by the nuclear reactions at the heart of stars. And it is this energy, in the form of heat and light, that powers the growth of life on earth (and presumably other places) through chemical reactions such as photosynthesis that store up the energy for controlled later use. Again and again we see a strong connection between particle physics (the very small) and cosmology (the very large).

Summary

Many physicists consider that these four forces may well be different aspects of a single phenomenon. In the same way, Maxwell showed that electricity and magnetism were two features of the one electromagnetic force and scientists are working to see if they can produce a single theory that covers all four forces. With the advent of quantum theory good progress has been made combining the strong, weak and electromagnetic forces into a single model but gravity has so far eluded our attempts to create a so-called theory of everything (or TOE).

All the forces apart from gravity are mediated by the exchange of subatomic particles (like photons) so this means they come in discrete quantum chunks. Gravity, however, is generated by the curvature of space-time and it is described by Einstein's relativity theory. Relativity and quantum theory are among the most successful creations of the twentieth century and they have both been experimentally tested and found accurate to incredible levels of precision. It is, however, proving very difficult to find a quantum description of gravity. The best hope so far is something called string theory but the maths involved are proving very complicated. Still, the search for a TOE is currently the holy grail of physics.

So, in summary, gravity, although the most familiar of the forces, is really pretty weird, it actually stretches both space and time. Electromagnetism is all about force fields that make empty things, such as you and I or the earth, appear solid. It is also responsible for light and chemistry (the interaction between atoms which we rely on, for example in digesting food). The two nuclear forces are responsible for how atoms are made and how they can change from one element to another, liberating vast amounts of energy. If any of these four forces were missing or even slightly different in some way we wouldn't be here at all, yet we mostly go through our lives without realising they even exist.

Chapter 6

Evolution

Not only did evolution happen: it eventually led to beings capable of comprehending the process, and even of comprehending the process by which they comprehend it.

Richard Dawkins

Evolution is a wonderful meme and an incredibly simple idea. Seldom in our history as a race, has one idea had such a profound effect on our thinking, our understanding and our behaviour. Having said that, however, there are quite a number of popular misconceptions as to its nature and its methods of working. In this chapter I want to discuss those false impressions as a way to developing a better understanding of evolution and of who we are and our place in the universe.

Before looking at those misconceptions though, let's start with a review of the basic concept. Simply stated, if an organism has what it takes to survive and to reproduce in the environment it finds itself in then it will leave descendants, and if they too are able to survive and reproduce then they will leave descendants. If the organism can't survive and reproduce then, obviously, it will not leave any descendants. Therefore, the only organisms we are left with in the world are those that are good at surviving and reproducing because anything else didn't leave any descendants.

So, we and all other currently living things (animals, plants, insects, bacteria etc) are descended from long and (crucially) unbroken lines of survivors.

How then do we get from pure survival to actual evolution (i.e. change over time)? Well, think about the population of a particular species. There will inevitably be variation in the makeup of its individuals, variation in size, shape, colour etc. For instance, there could be a mixture of eye colours: blue, green, brown and other shades in between. Some types of variation will be due to environmental factors such as whether they had a good diet or not when they were young but other differences will be genetic (like eye colour). It is the genetic differences that are passed on from parents to children via their DNA.

There are a number of ways that genetic differences can arise. When cells divide, they have to make copies of their DNA so that both the resultant cells have a complete set. This copying procedure is extremely accurate. It needs to be because if it wasn't the species would soon die out. However, the occasional error does creep in. Even then, cells have repair mechanisms that run through checking and mending corrupt DNA. Still though, the odd copying errors do occur and these could be anything from single letter nucleotide substitutions to whole gene portions being swapped around, deleted or even doubled up. As well as copying errors, there can be environmental damage to DNA such as chemical or radiation.

Given then that changes can occur in an organism's DNA, these will (only if they affect the sperm or egg cells) be passed on to the offspring. By far the majority of changes to a gene will either have no effect or will stop that gene from doing its job, which in most cases will severely handicap that offspring. If this handicap makes it less likely that the organism survives or reproduces then it will not leave descendants, as noted above. So again, the only organisms around today are ones where any changes to DNA were either neutral or (more rarely) beneficial.

There are exceptions to this. Each cell has two copies of its DNA so if one gene from the father say was not working it could just use the good copy from its mother (ignoring here asexual reproduction). Also, a change may not be totally debilitating but may

46

at the same time be useful in another way. Take for instance
cell anaemia, which is a disease characterised by misshapen red
blood cells. You might think that this would die out because it is not
good for survival but it actually gives protection from malaria so it
confers some benefit too.

The other important factor in evolution is the environment. It is
the local environment that the organism finds itself in that does the
sorting or testing of the genetic variations. Note that the environment
is not just the geography and climate of the area but it is also other
organisms such as predators, food or diseases caused by bacteria
or viruses. If a genetic change means the organism can't survive in
the environment it is in then it will not leave descendants. It is the
environment that does the shaping, in Darwin's words this is 'natural
selection'. He also used Herbert Spencer's term 'survival of the
fittest' by which he meant fittest for that **local** environment.

So here we have the requirements for evolution:
- A population with variation
- Mechanisms by which variation can occur
- A mechanism for passing on genetic information from
  parent to child (hereditary)
- A mechanism for selecting beneficial variations (i.e. the
  environment)
- Time for change to occur

Note in the fourth requirement the word 'beneficial'. It is
important to point out that this does not mean good in any abstract
or value judgement sense, only that it is of benefit to that organism's
survival and/or reproductive success at that point in time in that
local environment. *[I will expand on this below.]*

The fifth requirement of time for change to occur was initially a
big obstacle to the acceptance of evolution but as we saw in chapter
3, the biblical account of time was off by a huge factor. We measure
time by hours in a day, days in the week, months, seasons and
birthdays in the year. The length of time evolution has had to work
with is virtually unimaginable by our standards, a million years is just
a raindrop in the vast ocean time stretching back to the first stirrings
of life on the newly formed earth.

The idea of evolution had been around for a number of years

before Darwin but nobody had come up with a convincing mechanism for how it worked. Darwin did with his theory of natural selection and he did it without knowing anything about genes or DNA. He just reasoned, from the evidence he so carefully collected, that there must be some method of passing information about body structure from parent to child and, like all great ideas, in retrospect, it turned out to be very simple. As his great friend and supporter T H Huxley said at the time, 'How incredibly stupid of me not to have thought of that.'

Evolution Misconceptions:

The main aim of this chapter is to understand evolution better by correcting a number of popular misconceptions about its nature and how it works. Take the following statements:

- Evolution is a theory not a fact.
- They have never found the missing link.
- We evolved from monkeys.
- Evolution is about progress.
- Evolution is just random chance.
- Acquired characteristics can be inherited.
- Evolution is about perfection.
- Dinosaurs died out because they could not hack it.

Each one of these statements sounds reasonable doesn't it? You may well find them in the popular press or even expressed on TV. All of them, however, are wrong and they display a lack of understanding of the true nature of evolution. It is instructive to see why so I will discuss each statement in turn.

a) Fact or Theory

A common criticism of evolution (particularly by creationists or proponents of so-called intelligent design) is that it is just a theory. Not so. Evolution is a fact, it can be read straight from the fossil history in the rocks around us and it has been demonstrated time and again in biological studies around the world. One has only to think of the popularly named 'superbugs' currently invading our hospitals, against which the normal antibiotics don't work. These are bacteria that have **evolved** a resistance to our antibiotics.

Where the word 'theory' is actually used is not with the fact of evolution but with Darwin's proposed method of how evolution comes about, i.e. 'natural selection'. When scientists talk about Darwin's theory of evolution they are talking about the mechanism of natural selection, not whether evolution itself is true or not.

But even then, when we talk about Darwin's theory of natural selection, there is a further important distinction to make about the word 'theory' itself. When scientists use the word 'theory' they do so in a very rigorous manner, meaning 'a hypothesis that has been confirmed or established by observation or experiment and is accepted as accounting for known facts'. Note that this is very different to the colloquial use of the word in everyday language. Natural selection is a very well established theory (in the scientific sense). It does account for the known facts and has been confirmed many times by both observation and experiment, also it is virtually unanimously accepted by biologists throughout the world.

It is the confusion between the scientific use and the colloquial use of the word 'theory' and the lack of distinction between 'evolution' and 'natural selection' that leads people to say 'evolution is just a theory'. Anyone saying that is wrong on two counts and is expressing their lack of understanding of science in general.

Evolution itself, like the Big Bang described in chapter 4, has been demonstrated from a number of different, independent, sources:

Firstly, it has long been recognised from the physiology and anatomy of animals that they fall into different groupings, some more closely related than others ('related' being the operative word!). For instance, the domestic cat is more closely related to the lion than it is to the domestic dog and this can be shown by just looking at its skeleton.

A dolphin is a mammal and, as such, is more closely related to a horse than a fish, again its skeleton gives it away, one example being that the fish's tail moves side to side where as a dolphin's moves up and down, due to the mammalian str~~ucture of~~ backbone. Also of course, it gives birth to live you them, as the other mammals do. The fact that it looks like a fish is an example of convergent evolution, the oceanic environment dictating what forms and functions work best there. But inside, the dolphin's anatomy and physiology is a dead giveaway of its terrestrial origins, showing clearly that its ancestors moved back from the land to the sea.

Consider also Darwin's finches. When he visited the Galapagos Islands in the Pacific Darwin found many new species of animals. In particular he collected a number of specimens of the various birds that lived there. He initially thought they were all different, due to the very different beak shapes but when he got back to England and showed them to a bird specialist it was found that they all belonged to the finch family. This discovery aided Darwin greatly in his thinking about evolution. A pair of finches must have reached the islands in the past, probably blown off course by a storm. As they bred over the years different populations of them in different parts of the islands specialised their eating habits to different types of seeds and each population evolved a beak shape to deal best with their own type of seed. In fact, as these volcanic islands are relatively new geologically speaking (just a few million years old) it demonstrates just how fast evolution can work when it has the opportunity.

Then, of course, there is human anatomy, which is incredibly close to that of the other apes. The term 'ape' was invented to apply to chimpanzees, gorillas and orang utans but then it was found that we are more closely related to chimpanzees than chimpanzees are to gorillas, so if they are both apes we must be too. In fact back in 1758 (100 years before Darwin published his theory of evolution) Carl Linnaeus, the Swedish botanist who devised the Latin naming system for biology we still use today, put us, apes and monkeys in the order called primates. He was criticised for treating humans as animals but he challenged anyone to demonstrate any significant anatomical difference between us and the other primates. There were no takers!

Secondly, from the study of fossils we can trace the evolution of life right back to almost when life on earth began. Again, here our tools have helped because we have been able accurately to date when the different rocks were laid down and hence date the fossils within the different rock strata. Also, for the very earliest fossils, microscopes and electron microscopes have been invaluable. For more on fossils, however, see the next section, titled The Missing Link.

Thirdly, it was only in 1952, nearly 100 years after Darwin published *On the Origin of Species*, that Crick and Watson discovered the structure of DNA. Again, sophisticated tools were required to probe something as small as these tiny twisted threads of protein in the nucleus of cells. Every cell in our body has a complete set of all our DNA but once a cell has specialised and turned into a liver cell for instance, its control genes only turn on the genes needed for liver functions. There are around 25,000 genes in our DNA but this is only a small percentage of the total because there are large stretches of DNA that have no genes in it at all. This is the mechanism that Darwin deduced must exist to provide heredity.

Now we have the tools to analyse DNA samples taken from different animals and we can look at the differences. The first surprise was how few differences there were. But then that should not have come as such a surprise if we think about it because all cells have to do similar things such as building cell membranes and regulating the flow of chemicals and nutrients across those membranes. They also all need to create and store energy in some method. There are differences of course and scientists can measure these changes and so tell how closely related two animals are (i.e. how long ago they shared a common ancestor). This genetic analysis has been used to work out a family tree of life and has very much confirmed the tree constructed from using anatomy alone.

Fourthly, biology studies both in the laboratory and in the field have demonstrated evolution in action many times. Take the weaver bird for example. In its native South Africa it has a problem with cuckoos laying eggs in its nest and it deals with this by laying eggs with very specific patterns and colours so that the cuckoo's egg is very noticeable. These weaver birds were introduced into the Caribbean island of Hispaniola around 200 years ago and the Indian

Ocean island of Mauritius around 100 years ago. Neither island had cuckoos. The production of complicated egg patterns is costly in energy terms so since they were no longer needed the patterns gradually became less distinctive. Furthermore, in the Hispaniola birds, which have been isolated for longer, the fading is greater. This very nicely demonstrates evolution in action.

Many other studies like this in the field have been very well documented but scientists have also shown evolution in action in the laboratory. Fruit flies are perfect for this because their time between successive generations is so short and their DNA has also been fully sequenced. Many experiments have been performed with them. In one experiment, a few flies from a starting population were selected if they had a slightly stronger preference for seeking light than normal and then only those flies were bred together. In the next generation again only a few of those preferring light more strongly were selected for breeding, and so on for 20 generations. By this time the results were very marked with virtually all flies in the last generation showing a very strong desire to move towards light. This clearly showed that evolution can move very fast in one direction if the conditions are right.

Back in the field again, a team from Harvard University recently studied a particular species of lizard in the Bahamas. After the introduction of a predator lizard, the original lizard population soon came to consist mainly of individuals with longer than the former average leg length, obviously because they could run away from the predators faster. However, when the team went back six months later they found that most of the lizards had learnt to spend more time in the branches of shrubs to avoid the predators and these individuals had shorter than the former average leg length, for ease of climbing. Jonathan Losos of the team said it shows that 'natural selection can change direction on the drop of a dime'. It also shows again how quickly it can occur.

Fifthly, ideas from both maths and computing substantiate evolution. The great geneticist and mathematician Ronald Fisher, working in the early twentieth century, put evolution onto a firm mathematical basis. He worked out how genes should spread in a population and how things like the sex ratio (the number of males to females) in a species remains constant for that species. This gets interesting in species that have harems and in species like ants, termites and bees where there is a single queen surrounded by workers and fighters etc. Fisher's maths completed the picture, resolving a number of outstanding issues like the sex ratio that even Darwin had said would have to be left for future generations to work out.

Evolution has also, more recently, been used in computing. Instead of creating a computer program from scratch, some programmers have been able to evolve an algorithm to solve a particular problem. They initially generate a number of random programs and test them to see which ones come nearest to solving the problem in question. Then they take the best two or three and 'breed' a new population of programs from them by automatically swapping round bits of code and making small random changes. Then they test these again and select the best for breeding and so on for a number of generations. It is surprising just how quickly this produces an algorithm that solves the problem, indeed it is often a better one than would have been created if the programmer had coded it himself from scratch. This is essentially 'survival of the fittest' as in Darwin's natural selection, only in this case the problem to be solved was doing the selecting rather than survival in the local environment.

So evolution, verified from many independent sources, is most definitely a fact, indeed it is as much a fact as is the statement that the sun will come up tomorrow. Furthermore, Darwin's theory of natural selection is undoubtedly the mechanism by which evolution works.

b) The Missing Link

Again, this seems to be a criticism that creationists are fond of and this phrase has now entered into popular usage and gets applied to all sorts of situations. A number of points need to be made here. Firstly, the word 'the' in the phrase 'the missing link'. Is there just one fossil that, if found, would solve the problem forever? Some people do seem to be looking just for evidence that we evolved from apes, taking the (very odd) position that evolution is OK for other animals but not for humans!

The fossil record is notoriously patchy, it takes very special conditions for fossils to actually form and most dead creatures leave no trace at all. Add to that the fact that much fossil-bearing rock is underground or on top of mountains or even under the sea and you begin to see how lucky we are to have fossils at all. Having said that, in recent years and with much hard work and travelling to exotic locations, we do have a remarkable quantity of fossils to work with. Also, with the very diligent mapping of extremely convoluted terrain around the world and new tools providing accurate dates for the different fossil-bearing rock strata, we have been able to piece together a very convincing history of life on earth. Indeed a large number of so-called missing links have been found and identified throughout the historical fossil record. For example, whales are mammals that have returned to the sea from the land. Just recently, it was announced that a fossil ancestor of the whale had been found, it had stubby legs and lived by the sea shore. It was clearly an intermediary form between the ancestral mammal and the modern whale.

One problem with finding so-called missing links is the 'naming' one. Say such a fossil is found that is an intermediary form between an older ancestor and a more recent animal. Great, a missing link! But then this intermediary fossil will need a name because while it was alive it would have been in a species of its own right. After it has been given a name, there will then be two new gaps that require missing links, one between the older ancestor and the new fossil and the other between the new fossil and the more recent animal This process can go on and on with the sceptic not being happy until we have a fossil for every ancestor that ever lived for that species! Clearly an impossible task, so a pedant can always level the charge of missing link at evolution if he wants to, flying in the face of the enormous volume of fossil evidence we now possess that supports it.

If we return to the more recent timescales of human evolution (where the missing link criticism is most usually applied) again we have many fossils from within this period. However, even if we have the fossil of say a Homo erectus individual, that individual may well not be a direct ancestor of ours even if the species Homo erectus in general is in our ancestral line. So, it would be debatable whether to call that fossil a 'missing link' or just an example fossil of an ancestral species. In fact we have numerous fossils from this period but they just go to show how complicated a picture it is, with many species and subspecies, some on our ancestral path and some parallel to it.

We have Homo sapiens (us), H. erectus, H. habilis, Australopithecines robustus, A. boisei, A. africanus, A. afarensis and so on stretching back around two or three million years, all of which may be directly on our ancestral line or parallel to it. We also have Neanderthal, which was parallel and only died out around 28,000 years ago, although of course if you go back far enough we do share a common ancestor with Neanderthal. If some of these parallel species were still alive today instead of dying out it might not seem that the gap between us and our nearest cousins was as large as the currently perceived gap between us and chimpanzees. In fact, the same could be said for any two species, e.g. cat and dog, if all their parallel ancestral species had not died out we would have had many intermediate forms. It is the fact that most past species have died out that gives us a world today of quite separate animal species and challenges us when we think about the evolution of one species into another.

Why did our own parallel species die out? A number of possibilities have been put forward. It could have been climate change such as ice ages; competition for resources and land or even direct conflict with us or our ancestors. Indeed, many other large animal species also went extinct around the time that modern humans started spreading round the globe and a number of scientists see a direct link in this either through killing or habitat destruction. Species do go extinct all the time throughout the historical record, the average duration of a species is around 10 million years, although if you only consider land species the average is around 5 million years. However, we are currently living in a period of very fast extinction rates, comparable to the other five great mass extinctions in the past, like the one that killed the dinosaurs 65 million years ago and this current phase must, in no short measure, be due to our presence on the planet.

c) Monkeys on a Ladder

The first two invalid statements (fact or theory and the missing link) are more usually levelled against evolution by those who don't believe in it, although when you throw enough mud some does tend to stick and these two do seem to lodge in the public consciousness, casting some doubt where, if the facts were known, there is no need for scepticism. The following statements and ideas, however, (including 'we evolved from monkeys') are endemic even in supporters of evolution and need to be addressed if we are to properly understand the subject.

We have probably all seen the classic picture depicting evolution with a monkey on the left moving towards the right, next perhaps a chimp also moving right, then a bigger ape, then some sort of Neanderthal and lastly an upright human, all walking towards the right. Another image we have is that of a ladder of evolution again a single line leading up to humans as a kind of pinnacle at the top.

The main point to make here is that we did not evolve from monkeys. We and monkeys both share a common ancestor in the past that was not human and not monkey. Monkeys are at the current end of their branch of evolution just as we are on our branch. In fact, everything alive today is at the current end of its own branch and we share common ancestors with all of them. Instead of a line or ladder, a much better picture of evolution is a bush or a tree.

If we think about this further we see that to put humans at the top of some line or ladder is somewhat presumptuous and probably just a trick of perspective. OK we have a more complicated brain but that's just what we might think is important; other animals can run faster, giraffes may well consider that a long neck is really important. All other animals, plants, bacteria etc alive now are just as successful as us at surviving, they all have a continuous, unbroken line of ancestors stretching back into the past and they are all at the current tip of their evolutionary twig.

Note that many branches and twigs on our bush of life come to an abrupt end somewhere inside the bush because most species that ever lived are now extinct. We touched on this in the previous section when talking about parallel species. It is just the outside tips of the bush of life that are still growing and which have the potential for more branching in the future.

It is sometimes said that this is the age of man. Before that, we had the age of mammals, the age of the dinosaur, the age of fish perhaps and so on. Again, this is a trick of perspective. If you look at numbers of species currently in the world more than half of all animal species are types of beetle. They are an incredibly successful and adaptable type of animal. There is a story that the great biologist JBS Haldane once met a group of bishops on a train. One of them said to him something like, 'What do you deduce about the nature of God from all your studies of animals?' and Haldane replied, 'That he must have an inordinate fondness for beetles!'

But even more successful than beetles are the bacteria. They are incredibly resilient. They can be found in all environments on earth, from the poles to the tropics, from boiling hot springs to highly salty environments that would immediately kill any other species. They live on land, in the sea and even in the rocks far underground. It used to be thought that all energy for life on earth came from the sun but some species of bacteria have been found that live totally underground and obtain all their energy from chemical reactions in the rocks. Also, 10% of your body weight is made up of bacteria, mostly in your digestive tract and mostly very beneficial for you, aiding digestion and the absorption of vitamins etc. Indeed, because bacteria are about 100 times smaller than the typical human cell, you actually have about ten times more bacteria in or on you than you have your own cells!

However you want to measure it, number of individuals, number of species, total weight, number of different environments colonised, bacteria come out on top. This is really the age of bacteria, always has been, always will be. And they will still be here long after we have gone.

So, as we have seen, we did not evolve from monkeys, we and monkeys are both at the current end of our evolutionary twigs on the bush of life and we both share a common ancestor. Furthermore, we can make exactly the same statement about us and frogs, us and pine trees or us and amoeba, it is just that our most recent common ancestor in these cases is further back in time. Think bush not ladder.

d) Evolution and Progress

In general usage, the word evolution seems to carry very strong overtones of progress. Onwards and upwards, getting better all the time, improving and becoming brainier, with us no doubt at the top or pinnacle of the process. A bit like the ladder in the previous section.

This idea of progress seems to indicate a direction to evolution, even some point towards which it is moving, or aiming at. Nothing could be further from the truth! There is no way evolution could know anything about the future or how to aim at something there. All it deals with is the present, that which works best here and now in the local environment.

The notion of progress is a Victorian one, from a time when society was becoming industrialised and things seemed to be getting better, striding confidently into the future. And as evolution became accepted, it naturally seemed as if there was a progression from amoeba through fish, reptile and mammal up to man as the most highly evolved thing on the planet. However, all the word 'evolve' really means is 'change'. It is true that sometimes there are selection pressures in particular directions (I'll mention these in more detail below) but the only thing natural selection has to work with is the local environment.

Another problem with the concept of progress is how do you measure it? Are we just talking about braininess? Because as we saw in the last section we might as easily talk about long necks, speed of running or even size of animal. One attribute we might accept could be some measure of the complexity of an organism, surely complexity has increased over the millions of years since life began. Well, yes and no. There are more complex life forms now but the simpler ones are still with us. As we saw in the previous section this is still the age of bacteria and they are not getting brainier or more complex, neither is grass but it is doing very nicely at the current end of its own evolutionary twig. In fact, if you look at any animal, giraffes, mice, fish, or beetles none of them are getting brainier or even more complex. There is no in-built direction towards braininess or complexity in evolution. In fact, evolution can go both ways, functions that have evolved in the past can be lost again. This has happened many times in evolution.

The obvious example is parasites. This is where an organism that previously foraged and survived by itself changes to live off another organism. Often it loses functions it previously had, in some extreme cases all that is left is the ability to feed off the host and reproduce. Again, it is at the current end of its evolutionary twig on the bush of life but you would not call it complex by any means.

Others (non parasites) have also lost functions. Take for example the Mexican cave fish; they are closely related to relatives that live outside the caves but something made them move into the caves and start living there. The fish in the caves have now lost the function of sight, obviously it was of no use to them. Not only that, they have also lost their skin colourations, again no use in their new dark environment. If any of their relatives outside the cave had a mutation that caused their eyes to get worse or even stop functioning altogether that fish would not last too long but in the darkness such a change made no difference. Evolution preserves best that which is most useful for survival and reproduction.

We can actually see this when we look at DNA and compare similar DNA stretches between different animals. As you would expect, in animals that are more closely related, the similar stretch of DNA shows fewer changes than in animals that are more distantly related. However, if a particular stretch of DNA is vital for some specific purpose (its involvement in a particular chemical cycle for instance that is needed for cell function and can't easily be done any other way) then even in distantly related animals that DNA stretch will have very few accumulated differences. Scientists can measure these differences for particular stretches of DNA and can calibrate how many should occur over a known time period. This is a very useful tool for identifying exactly when and in what order species diversified from each other, hence enabling us to map the evolutionary bush of life.

There are many other examples of loss of function, for instance another cave dweller, a particular spider in caves in Australia, also lost its sight whereas its relatives outside the cave actually have eight eyes! Also the ability to fly has been lost, not once but numerous times. The ostrich obviously got big enough and fast enough at running that it didn't need to fly to escape predators. Other birds lost the ability to fly when they got to islands where there were no natural predators. A famous example here was the dodo, which was a relative of the pigeon and which reached the island of Mauritius where, with no natural predators, it lost its ability to fly. It also lost its fear of large animals and when the Portuguese sailors arrived in 1507, the dodos were extremely tame. Even though they were not that good to eat they seemed to have been clubbed to death just for sport – no wonder they soon became extinct.

Another example of loss of function is right behind you. Our coccyx at the base of our spine is all that is left of the time when our ancestors had tails.

So, in terms of complexity, evolution can go both ways. Let's say left is more simple and right is more complex. Now there is an obvious left-hand boundary here, life can't get too simple or it would not be able to continue existing but it must have started at this left-hand wall so it starts to move to the right. At any given time it can move left or right so, over time, there will be some species that become more complex but there will be many that either stay where they are or become less complex. This is why I answered 'yes and no' to the question 'Has complexity increased since life began?' Yes, there are more complex organisms in the world today but this is inevitable given plenty of time, a left-hand wall and the fact that evolution can go both ways.

I have just tested this on the computer. I wrote a simple 14-line program in excel. First I set 50 cells in column A to the value 1, to represent the lowest possible complexity (higher numbers will represent higher complexity). The program, in each pass, ran through each of these cells and, at random, either added 1 to its value, left it alone or subtracted 1 (a change of 1 being the smallest possible change in complexity, up or down). The one constraint being that the lowest possible value was restricted to 1 (the left-hand wall of least complexity). After the program had run for 100 generations, the results were quite dramatic. There was a whole mixture of values, mostly grouped around the lower end but at the other end there were two 17s, an 18, a 19 and a 23. Complexity appeared with no in-built direction to the changes! And that was just 100 generations, what could be done with hundreds of millions!

So, is there an inherent direction in evolution towards complexity? No! As clearly demonstrated by the fact that most organisms in the world today are not very complex, even though they have evolved for exactly the same length of time that we have.

I said earlier that sometimes there are selection pressures in particular directions. Well there are but even then, it is not always in the direction of complexity. You often get an arms race situation developing, say between host and parasite or between predator and prey. So for instance an antelope that has genes which enable it to run faster will survive better and be more likely to reproduce, hence its children will also have this ability. But then only lions that can still catch them will survive and go on to breed so the lions will get faster too. Such a selection pressure continues in the initial direction until often a plateau is reached where it is not cost effective (in evolutionary terms) for it to continue. This is where the phrase 'survival of the fittest' comes in. We might also remember Tennyson's line in his poem 'In Memoriam' which talks about 'nature red in tooth and claw'. Evolution is extremely ruthless in weeding out anything that is not fit for the environment it is in. Of course, that is not to imply any morality here, one way or the other, evolution is neither intrinsically good or bad, it's just natural.

Another selection pressure is sexual selection. If a female of a species finds (for whatever reason) something particular in a male that is attractive, say a red beak, and mates with him, then their offspring will carry both the genes of the female which could include the red beak fancying ones and the genes of the male coding for the red beak. Any males with reddish beaks would start to be selected over the other males and over time the average male beak colour would get redder and redder. Many birds are incredibly brightly coloured and since they don't need colour for camouflage because they can just fly away from danger, this is often an example of runaway sexual selection.

A perfect example of this is the tail of the peacock, which is incredibly decorative, whereas the peahen's tail is quite drab in comparison. In this case it is actually quite costly to the male to produce such a display so in choosing a male with a good tail the peahen is also probably choosing a fit male with few parasites who can afford to put a lot of energy into producing a good display, and if her eggs hatch as male they too should have good displays ensuring they get chosen by future peahens, a good way of ensuring the survival of her genes, a win-win situation.

And what of our own appreciation of beauty? We would probably rate a person as more attractive if they were symmetrical, youthful, vigorous and, in the case of a female, perhaps slightly curvaceous, all indicators of good breeding stock and lack of disease. Are our own values shaped by evolution as well? Certainly, evolution must have had some effect on the way we think and it can only add to our understanding if we acknowledge the possibility. *[See chapter 10 for more on this.]*

Another example of change in a given direction is where we have deliberately selected for it in our domesticated animals and plants. Like the fruit fly I mentioned earlier, if selection is guided in a given direction things can move very quickly. For instance cows selected for milk or meat, plants selected for increased yield, pigeons for racing etc. A particularly dramatic example is that of dogs. All current species are descended from just a few wolves (confirmed recently by DNA analysis) and bred for whatever trait people wanted: size, obedience, docility etc. Every type of dog from Alsatian to Pekinese was created in just a few generations.

Evolution is also confused with the idea of progress because it is sometimes used that way in non-biological terms. Astronomers sometimes talk about the evolution of a star, by which they mean over the lifetime of a star it might go through different phases from its birth out of a collapsing dust cloud through a period where it is like our sun, then on to a red giant and a white dwarf. The point here is that astronomers are talking about an individual star going through different phases over its lifetime whereas in biology an individual organism does not evolve itself, it is species that evolve and, as we have seen, the direction is not predictable.

When the concept of evolution was confirmed and brought to general attention in the mid to late nineteenth century by Darwin and his supporters, the idea was quickly confused with that of progress. Victorian society was seen as a pinnacle of civilisation and the fact that tribes of 'savages' were being discovered in Africa and other places only seemed to confirm the idea of progressive evolution. The concept was applied in as many areas as people could think of. Karl Marx even picked up on it in his writings and saw society and government 'evolving' through different levels 'up' to communism as the highest and best form of government.

Of course, the biggest misuse of the idea of evolution was by Hitler. He saw the 'Arian race' as the pinnacle and indulged in massive programs of ethnic cleansing to promote his ideas. He saw tall, blond, blue-eyed men as the natural rulers of Europe and probably the world, an incredible misuse of the theory of evolution mixed with the idea of progress.

As we have seen, we are not currently getting brainier, indeed we have hardly changed over the last 200,000 years. So-called 'savages', living in jungles, are genetically exactly the same as us, just as intelligent and with their own store of knowledge about their own environment (which Westerners would probably have a hard time surviving in). The difference is one of memes; ideas and culture. *[See chapter 9 on culture later in the book.]*

So the idea that evolution is about progress is not only wrong, it is dangerous. We have, by incredible good fortune, tremendous capacities for rational thinking, creativity, self-awareness and an appreciation of beauty but we would be foolish to see ourselves as some sort of inevitable pinnacle of creation. We could disappear tomorrow quite easily and the world would go cycling on for many more millions of years quite happily without us. While we are here, however, it is incumbent upon us to use our gifts wisely for the stewardship of the planet and the wonderful variety of life that has evolved alongside us.

e) Random Chance

You might hear the phrase that 'evolution is just random chance'. Or you may hear someone say, 'I can't believe that we evolved just by chance.' The speaker may be denying evolution altogether or, more subtly, they may accept evolution in some form but still think that it is too much for chance alone to do, so they postulate some other 'force' at work as well or maybe even a guiding hand in some form. The mistake here is in thinking that the theory of evolution is about random chance. It is not. It is about natural selection. If it were all about chance, it might be like a hurricane blowing through a scrap yard and assembling a jumbo jet. Even if the hurricane blew for billions of years it would not succeed, simply because a) there is no heredity and b) there is no selection.

Random chance comes into the picture in two ways. Firstly, natural selection can select only when there are variations to choose from. In a given species, if some members of that species had say thicker hair on their skin they might survive better in a coming ice age. As the ice got closer, thicker and thicker hair would probably be selected for (in this case by way of those members with less hair dying and hence not leaving descendants). But where did the variation come from? As we saw before it comes from things like copying errors when cells divide or mutations caused by radiation or chemicals or the like. It is the generation of variation that is random. Most of these genetic variations are either neutral or positively harmful to the individual but the odd one or two, like a gene variant that gives thicker hair, could be useful. Given then that the origin of variation is random, natural selection definitely is not. Things look as if they have been designed because they have been selected by the environment, or, more accurately, anything that didn't work was selected against.

The second way random chance occurs is via catastrophes. This may be as simple as an accident where an otherwise fit individual died before passing on their genes. In this case others of the species would also have that individual's genes so they would probably make it into the future, but there is always the possibility that that individual had a particular variation that was unique and then it would be lost altogether. More seriously though, there may be a catastrophe (like a flood, an ice age, a disease or an asteroid impact) that wipes out a whole species. Species have gone extinct in the past through no fault of their own, in fact many are going extinct now, mainly due to human activity.

So, evolution is not random and it is not guided by any outside force; natural selection is the driving force, the 'designer'. It is a ruthless worker but what wonders it has created here on earth and, most probably, elsewhere in the universe!

f) Acquired Characteristics

The idea of evolution had been around for a good number of years before Darwin (it was obvious from anatomy that some species were more closely related than others) but nobody had worked out how new species could occur. There were a number of possible theories around and one in particular was championed by the French biologist Lamarck who in 1809 published his theory of evolution by the inheritance of acquired characteristics. This theory proposed that if organisms or animals gained some characteristic during their life then this addition would be passed on to their descendants. This would be like a blacksmith working at a forge with his hammer all day and developing a big, strong right arm and then passing that on to his children so they had a strong right arm even if they didn't use it as the blacksmith had.

We now know that evolution doesn't work that way. If it did then it could move at a much faster pace than we know it actually does. There is no mechanism for passing learned characteristics to descendants because they inherit just the genetic DNA from their parents' egg and sperm cells and acquired characteristics do not modify the DNA. Of course, they knew nothing of DNA in the nineteenth century but Darwin worked out that Lamarck was wrong through logic, experiment, observation and deduction. Science at its best. It is a pity though that Lamarck is remembered today mainly for his incorrect theory of biological evolution because he was in his day an excellent biologist and was the curator of the French Natural History Museum.

It would be wrong to think about genes as things that completely determine the nature of a person, like say giving someone a big right arm. What they provide is the potential to develop something. They provide the capability for growth (of the arm muscles say) given the right environment. The blacksmith's children could themselves develop a strong arm if they used it as much as the blacksmith, so could most other people.

Genes are not a blueprint for building a person, a more accurate simile is that they are like a recipe in a cookbook. They provide instruction for using the basic ingredients when making a cake but how it turns out depends very much on the environment it was made in, e.g. the shape of the tin, the temperature of the oven, the humidity, how the ingredients were mixed, indeed all sorts of things. Genes provide the potential but how this is realised is very dependant on the environment.

There is one area, however, where the Lamarckian style of transmission does work and that is culture. If you learn how to do a particular thing, like say making an axe, you can show your children how to do it. One of them might come up with improvements to the method and then show others. This is what I talked about as memes in chapter 2 and this method of spreading ideas is extremely rapid and (unlike Darwinian evolution) it can have a direction towards improvement. *[For more on this see chapter 9 on culture.]*

g) Are We Perfect?

There is a widespread belief that evolution is about the perfection of design. We think of the tiger perhaps as the perfect killing machine or the thoroughbred stallion as the ultimate running machine. Again, this idea is strongly tied up with the concept of progress in evolution. If we take the horse for example, this design is not that successful, indeed it is quite cumbersome for anything other than running. In fact, if we look at the fossil record, horses seem to be on their way out because there is only one species of them left (plus a handful of species of zebra, donkeys and asses). Ten million years ago there were around 16 species of horses, better but certainly not as successful a body plan as the beetles (or even rats, bats and antelopes, which are the most successful of the mammals).

Then look at humans, are we that well designed? We certainly have many problems with things like knees and backs, mainly due to our relatively recent evolutionary change to a more upright posture. Also, many of us need glasses to correct our vision. This, by the way, pointedly belies the concept of intelligent design, a good engineer would never have come up with these solutions. The point here is that an engineer would be starting from scratch and could also borrow good ideas from others, whereas evolution can work only with what is already there. Also, every intermediate step has to be a fully working design in its own right so evolution is very constrained as to what it can do.

More pertinent than this, however, is the fact that evolution has no direction, no knowledge of the future and nothing to aim at other than survival and reproduction in the local environment. Any improvements it might make to a design, while maybe helping in survival, are often costly in terms of other things like energy requirements. In fact, if a feature can be lost without affecting the prospects of survival and reproduction then it probably will disappear.

I am reminded of the story of a multinational car manufacturer that built a large test track complex to try out new cars and modifications to existing models. One of the things it used the track for was to hammer a car round and round to see if a particular component could be held on with just two bolts instead of three. Evolution does the same, it tries to get by with the least possible. It is a make-do approach, constrained by having to work with existing designs; it is definitely not about perfecting design, it does just enough to get by.

Having said that, some things, like for instance the eye, have been improved by many changes, enhancements that deal with problems like spherical aberration, chromatic aberration, varying light intensity, focus etc. In these cases the cost of the changes are outweighed by the advantages of being able to locate food, find mates and avoid predators more easily. But even here, evolution does the least possible to get by. We saw how the fish that moved into the darkness of caves soon lost their eyes. Also not all animals have colour vision, presumably what they do have is enough for them (remember, they too are at the current tip of their evolutionary twig on the bush of life). Indeed there are a number of different eye designs, for example the compound eyes of some insects, all doing just enough for the animal involved.

Perfection, like progress, is an illusion generated by natural selection working over billions of years of deep time. Improvements do occur over time, sometimes even surprisingly quickly, but these are just the least costly solutions to local problems, not some general movement towards perfection by any means.

h) Dinosaurs Couldn't Hack it

There is a common notion that dinosaurs were lumbering, slow-witted creatures that lacked some quality that would have enabled them to survive. You sometimes hear the word used in a derogatory sense, perhaps with overtones of pity, when someone says, 'he's a bit of a dinosaur' meaning old fashioned, stuck in his ways, soon to be extinct. Indeed just recently a TV newsreader was interviewing a creationist and the newsreader, in support of evolution, expressed the view that humans evolved after the dinosaurs died out because (and I quote) 'the dinosaurs could not hack it'.

Nothing could be further from the truth. The dinosaurs were incredibly successful animals and they dominated the earth for over 180 million years. We would be lucky if our tenure on this planet lasted anywhere near as long. What actually brought an end to the reign of the dinosaurs was a global wide catastrophe 65 million years ago, most likely in the form of a large meteor. This not only finished the dinosaurs, it took out most other species on the planet at that time, even including many species of marine life. Our ancestors made it through only because at that time they were small burrowing mammals that existed as scavengers. They later evolved to fill the niches left by the demise of the dinosaurs and other large animals that perished in the catastrophe. If that meteor had not arrived, we would not be here now – such is the element of chance in evolution.

The earth is hit by meteors all the time, luckily most of them are small, but ones the size of the meteor that killed the dinosaurs hit, on average, once every 100 million years or so. There is no doubt that, some day, another one of that size will hit the planet again and if humans are still here at that time they would probably not survive the devastation such an impact would cause, leaving perhaps other small burrowing animals or insects to repopulate the earth. In fact there are also a number of other possible causes of such global catastrophes, including massive ice ages covering not just part of the northern hemisphere but most of the planet (it has happened in the past); huge volcanic events blanketing the earth in poisonous gasses and blocking the sun for years so plants could not grow and even nearby supernova which would flood radiation across the world. Such global catastrophes, although rare, have occurred on earth at least five times in its history and, in the long term, another mass extinction event is inevitable.

So, the dinosaurs very much could 'hack it'. There were many species of them from small quick runners to magnificent large grazers and of course the top predator Tyrannosaurs Rex. They ruled earth for a long time and only the random movement of celestial mechanics brought their dominance to an end. Having said that, there is one relative of the dinosaurs that survived the catastrophe and is still alive today and doing very nicely thank you. That is the birds! They are the closest living relatives to the dinosaurs and they evolved directly from one species of dinosaur into all the different birds we see today. The dinosaur is definitely one of nature's success stories.

Summary

You sometimes hear the argument that says, I can't believe that something as complex as a person (or even just an eye) could evolve from a primitive single cell organism just by random chance.

Well I've dealt with random chance above but the real problem here is in looking at and being daunted by the huge gulf between a person and a single cell organism. Dawkins provides an excellent simile here, he says it is like climbing Mount Improbable. From this side, just look at those huge cliffs we have to get up, it must be impossible. But if you go round the other side, we find that there is a long but gentle slope all the way up to the top. All you have to do is put one foot in front of the other in small steps. It is quite easy to imagine small changes, perhaps say from Homo habilis to Homo erectus. All you need is many millions of such steps, and it was the discovery of deep time that made this understanding possible. The three and a half to four billion years since life began on earth is more than enough time for evolution to work its wonders and create the incredibly complex and interrelated natural world we see today.

Even between us and other apes the differences look daunting but that is mainly because the intervening species (like Homo erectus) between us and our common ancestor with the other apes no longer exist today. Another point to make here, however, is that you can get quite big differences with just small genetic changes. That is because many of our genes are control agents, i.e. they switch other genes on and off at different times in our development. If you think of say a millipede, much of its body length is just repeated segments with a bit of body and two legs attached. Now there will be a set of genes that work to build a single body segment but there will also be controlling genes that determine how many segments to build. A small change in such a controlling gene could produce more body segments, quite a large effect for one small change.

If you look closely at chimpanzees, you might see that the faces of younger chimps actually look more human in structure than the adult chimps do. Now, if a change happened to the genes that control the speed of development, such that the young chimp doesn't turn into an adult for a much longer time, then its youthful growing phase could continue for years more than usual. Instead of stopping when it normally would its body and its brain size could continue growing for longer. Indeed humans do have an extremely long childhood development period compared to the other apes. This process of

the slowing down of maturity is called neoteny. A
one or two controlling genes and a big change re
gap between us and our nearest relatives does
especially when you consider that evolution has ha
years to make these changes since the time we last sha
ancestor with chimpanzees.

In his book *Natural Theology,* in 1802, the reverend William
Paley famously said that if he was out on a walk and came across
a rock on the ground, he would consider it natural but if he came
across a watch he would see immediately that it was different to
the rock. He would be prompted to ask questions like 'How was it
made?, 'How does it work?' 'Who designed it?' because it would
obviously look as if it had been designed. Paley gave this argument
of design as evidence that there must be a god.

I would take slight exception to the first part about the rock. This
book is about seeing beyond the seemingly obvious or self-evident
face value of things. As we have seen in the previous chapters, if
we use other tools than just our eyes we can investigate the rock in
detail, What it is made of, how it was formed, what is its history and
then deeper still into its constitute atoms etc. A whole universe in a
rock.

Having said that, I do take the point that the watch looks designed
and that this begs the question 'Who designed it?' And then, by
analogy, we look designed so who designed us? There are of course
a couple of important differences between a watch and a living
thing. Firstly, the watch doesn't contain within it the instructions for
its own growth (DNA) and secondly the watch doesn't reproduce
to leave descendants. Now you might think that this makes living
things even more complicated than a watch but it is exactly these
features that enable evolution to work. Darwin gave us the answer
to the question 'We look designed so who designed us?' Natural
selection is the designer. In Dawkins' words, natural selection is the
blind watchmaker.

One final point to make on Paley's argument is that in the final
analysis it is self-defeating, even without knowing anything about
evolution. If something is so complex that by definition it must have

signer, then surely that designer must be even more complex so
 by the same definition he too must have been designed; the argument is exactly the same and could go on ad-infinitum.

The current fad of intelligent design runs into exactly the same
problem when proponents say that such and such a biological process
is too complex to have evolved therefore it must have had a designer
of some sort. It is an answer that says nothing and instead of using
science to investigate things, it is reverting to the cop-out of 'magic'
to provide answers.

I'll finish this chapter with a quick look at what has been a
fascinating debate within the biology community over recent years.
That is, where exactly does evolution occur and what is it that gets
selected by natural selection?

The obvious answer to that question is that it is the individual
that gets selected because, if he dies, without leaving descendants,
then he is not selected. However, a case could be made that evolution
happens at the species level, say where a population becomes
divided, by a mountain range for instance, and over the years the
two groups drift apart genetically until they are unable to interbreed
even if they did meet again. Then they would be considered two
distinct species.

A more controversial idea though, and one that has developed a
good deal of credence over recent years, is that selection happens
at the genetic level. Again, one of the main proponents of this view
is Richard Dawkins. His first book *The Selfish Gene* sets out this
position and most of his other books since then have expanded on it.

It is the actual genes themselves that are changed when copying
errors or mutations occur. When a gene is turned on in a cell and
starts working it will have a particular effect on the animal (like
producing hair of a certain thickness). Such physical or biochemical
characteristics are called the phenotypes of the animal. Some
changes to DNA do not alter the animal's phenotype and hence
there will be nothing for selection to work with but as soon as there
is a change that affects survival or reproductive success then natural
selection will work on it.

Successful genes will spread through the gene pool of a species and it is the genes themselves that flow down the years into the future. It is almost as if individuals within a species are just the temporary containers for that species' genes and you could say that a body is a survival machine that genes build, in which they meet and interact with other genes. Also, because genes get shuffled and mixed at the point of conception the strange possibility exists that someone could be your physical ancestor but you actually carry none of their genes.

Within the gene pool of a species, many of the genes will exist in a number of slightly different variations. It is these variations that natural selection works with. Some of the most successful changes are where genes work together so that if one gets selected then the other one will too by default, other genes in the pool being part of the environment that individual genes have to cope with. Here though, some changes might get themselves selected by being associated with useful genes even when they are not useful in themselves.

This view of evolution as acting at the genetic level is very persuasive, although one must be very careful to remember that when talking about 'selfish' genes or even when saying evolution does this or that, we are not implying any forethought or motivation to these blind processes. So where does evolution act, at the genetic level, on the individual or on the species? To some extent it is a matter of viewpoint and it is interesting and even useful to be able to look from each level at different times but in the final analysis it is the genetic level where changes actually occur and it is those changes that actually get selected for via their phenotype.

Some see this whole vision of evolution as harsh and uncaring, indeed this is often their main criticism, so I'll leave the final words in this chapter to Darwin himself: –

From so simple a beginning endless forms most beautiful and most wonderful have been, and are being, evolved... There is grandeur in this view of life.

Charles Darwin

Chapter 7

Life's History

Stop and consider! Life is but a day;
A fragile dew-drop on its perilous way

John Keats

Let's return to our picture of deep time and try and put some milestones in place. In chapter 3 we had the image of those few fingernail filings floating to the floor representing the entire span of human existence on earth but what about other events in the history of life, when did they happen?

Another way to visualize deep time is to think of a long piece of paper, 4.6 m long in fact. Imagine that the bottom of it is touching the floor of your living room. Assuming your rooms are about 2.4 m high (around 8 ft) the paper would stretch up through the ceiling into your bedroom above and, allowing say 20 cm for the thickness of the floor, the top of the paper would be just 40 cm from your bedroom ceiling. Now think that the paper is marked out in millimetres along its whole length. If each millimetre represents one million years, then the full length would represent the history of earth since its creation at the birth of the solar system, at your living room floor, up to the present day, near your bedroom ceiling.

Take the very top millimetre, near your bedroom ceiling and divide it into five equal parts, then just take the top fifth of that top millimetre, this would represent the whole history of the human race on earth. One fifth of 1 mm! A minute fraction compared to the rest of the history of earth stretching all the way back down, through the bedroom floor and down to the floor of the living room below.

Now I don't want to give the wrong impression with this paper history time-line stretching up through the ceiling. Remember the discussion in chapter 6, evolution is a bush not a ladder and it has no inbuilt direction of progress. This length of paper is just a way to visualise the timescale involved and to put down markers as to when different levels of complexity appeared.

The oldest rocks we know on earth are from Isua in Greenland and date from about 3.8 billion years ago. These rocks are sedimentary and hence must have formed from the erosion of even earlier rocks long since wiped from earth's history by the ravages of deep time. Even these rocks bear the chemical traces of early life. Much before this time the earth was still being bombarded frequently by the debris left over from the formation of the solar system. So, it looks as if life in some form started almost as soon as it was possible to do so. This marker on our paper would be 0.8 m (2.5 ft) up from your living room floor.

The first single-celled microbes appeared about 3.6 billion years ago, 1 m up from your living room floor. These were very simple with no central nucleus and with its DNA spread round the cell. We then get a long gap before the more complex cells with a nucleus to hold its DNA appear at 2.4 billion years ago. This would be just 20 cm below your living room ceiling, that's almost half the age of the earth just to form more complex single cells. Just before this time the important chemical reaction of photosynthesis emerged and this allowed atmospheric oxygen levels to rise, which in turn gradually allowed more complex chemical reactions to evolve.

The first multicelled organisms didn't appear until about 1.4 billion years ago, well into your bedroom, say 0.6 m (about 2 ft) above your bedroom floor (assume the floor thickness to be 20 cm).

Then around 540 million years ago what is known as the Cambrian explosion of life occurred, a fairly sudden appearance in the fossil record of many different types of marine animal. This would be about 1.46 m (nearly 5 ft) above your bedroom floor. Their sudden appearance as fossils is mainly due to bone and shell evolving, which leaves a much better fossil record than soft tissue.

So, we are now just 54 cm (21 in) below the top of the paper and only now have the first real animals appeared and they are still in the sea! The first real land plants can be marked at 43 cm from the top, the first crawling land animals at 41.8 cm from the top and the first insects at 40.4 cm from the top. The first reptile appeared around 32.7 cm from the top, mammal-like reptiles at 28.4 cm from the top, the first dinosaur at about 22.6 cm from the top and the first mammal at just 20.9 cm from the top.

The age of the dinosaurs lasted from 22.6 cm from the top to just 6.5 cm from the top, when they were killed off in the most recent of the five great mass extinctions the earth has known 65 million years ago. At this time, our ancestors were small burrowing mammals who were lucky to survive the catastrophe, just 65 mm (2.5 in) from the top of our length of paper near your bedroom ceiling! We evolved from those small burrowing mammals in just the last 65 mm, a tiny fraction of the time-line of earth. We shared a common ancestor with chimpanzees at just 6 mm from the top and, as we saw at the start of this chapter, humans as a race emerged in Africa around 200,000 years ago, the top fifth of the top millimetre!

Looking at the timescale it seems very much as if evolution has become quicker over time. It took a long time for the first single-celled organisms to arrive, then another long time for the evolution of multicellular life and then for simple creatures to appear. Then it seems to speed up with more and more variety and with us evolving from a common ancestor with chimps in just six million years (the last 6 mm, very short geologically speaking). This speed up, however, is a bit of an illusion, the reality is that the first part was so difficult, almost half the length of time the earth has existed was needed for complex single cells to appear.

Let's try a different picture. Imagine you have been given a box of children's Lego building bricks. It would probably contain bricks of different sizes, windows, doors, roof tiles, even wheels, gears and the like; you could very quickly build all sorts of fantastic structures. But now imagine that you are given a box of raw materials and told to first make the bricks from scratch. It would take you a very long time. It is the same for the first simple-celled organisms. Cells are very complex things, based on multiple chemical reactions that do things like creating energy, storing it and then utilising it in other chemical reactions to control all the processes a cell needs. Things like the transportation of chemicals, molecules, proteins and nutrients both ways through the cell wall, building the cell wall in the first place, repairing itself when damaged, constructing proteins from basic amino acids and reproducing itself. Then, once you start to get multicelled organisms some cells can start to specialise so they can better work together as a community (the doors and windows of the Lego set).

I don't want to stretch the Lego analogy too far, you can't just assemble a new species from scratch. In Lego terms the creation of a new model would be more like taking an existing one and changing just a few bricks one at a time, say changing a wall brick for a window one. The main point with the Lego comparison is the difficulty in creating the basic components in the first place.

So evolution has not really speeded up, it is just that it was very hard to start with to develop all the initial chemical reactions needed to support complex cells. Indeed when we look at our DNA we find that most of our genes are involved in the construction and maintenance of our cells, one reason we share so many of our genes with other animals and even with plants. Many of these genes are so crucial to basic cell metabolism that they have to be preserved virtually unchanged or they would stop working altogether.

Before we get even the simplest of cells, however, the earliest stirrings of life would have been just chemical reactions and self-reproducing strands of RNA (which is a precursor to DNA) but where did these come from? In 1953 the chemists S L Miller and H C Urey first performed what is now a famous series of experiments. They assembled in a jar a mixture of methane, ammonia, hydrogen and steam (which is assumed to be the composition of the earth's early atmosphere) and then they passed electrical discharges through it to mimic lightning. The result was the build-up of a rich brew of organic chemicals including many amino acids. These are the building blocks of proteins and nucleic acids like RNA and DNA, so it seems very likely that the basic ingredients of life were probably being created right from the earliest days of earth's history.

Further chemical reactions would naturally occur, with simple organic molecules joining together to make longer organic molecular chains and, with millions of years to brew, all sorts of different combinations would be tried. It only needed one reaction to become self-sustaining out of the billions that would have occurred throughout deep time and life would get a foothold.

As well as the creation of organic molecules on earth through basic chemistry and lightning, another theory of organic synthesis, called Panspermia, is gaining credence. By this theory, organic molecules are created in space in clouds of dust and gas or on the surface of icy comets, with both of these places getting zapped by cosmic rays and starlight. Astronomers have indeed detected organic molecules in space by studying how the spectra of stars are changed by intervening gas clouds. Different molecules absorb light at different wavelengths in distinctive patterns, easily betraying their existence to anyone with the correct tools and the patience to look. So these organic molecules could have rained down from space itself or they could have arrived on the many icy comets which hit the early earth. There is of course nothing to prevent a mixture of both this and the lightning theory being correct.

Photosynthesis evolved around 2.8 billion years ago, from then onwards the atmosphere started to contain free oxygen but before the Cambrian era, which began about 540 million years ago the concentration of oxygen was still too small to support multicellular animals. At the start of the Cambrian era something caused a dramatic rise in the amount of free oxygen. Carbon reacts with oxygen and this usually removes it from the atmosphere but as more and more simple life appeared much of earth's carbon started to get locked up and buried as the remains of dead organisms, increasing the amount of oxygen available which allowed more complex life to be sustained in a slow building feedback loop.

The appearance of more complex abilities throughout the history of life has by no means been a smooth incline. There were long periods of no increase and then sudden advances which themselves triggered other adaptations. Take the eye for instance. One of the first creatures to develop eyes was the trilobite about 543 million years ago. This was quite a sudden development in terms of geological timescales because its trilobite ancestor of 544 million years ago did not have eyes. These eyes were compound ones, similar to those of modern insects. It has been calculated using a computer simulation that it is possible to evolve eyes from an initial area of light sensitive skin in just half a million years.

Interestingly, as we saw earlier, this was around the time of the Cambrian explosion, when hard bodied, well-protected animals first appeared in the fossil record. When a predator like a trilobite can see its prey then the only ones to survive are ones that can protect themselves with shell and bone or, if they too develop eyes to see the trilobite coming, they can run and hide. Other animals did indeed develop eyes around this time too. In fact, there was suddenly quite an arms race going on. So, the evolution of the eye may well have been the cause of the so-called Cambrian explosion.

Although the original prototype eye probably only developed once, earlier in history, by the beginning of the Cambrian a number of different species had these early receptors and began to evolve their own eyes. There is quite a large variety of different types of eye (possibly as many as 60+ variations on around five or six basic design principles) and many of these evolved separately into fully working eyes very quickly around that time. Hand in hand with eye development was brain-processing power. The images provided by the eye are useless unless they can be interpreted and then acted upon. A very large part of our brain is devoted to the interpretation of visual perception. *[See chapter 10 on evolutionary psychology.]*

When Darwin first put forward the concept of natural selection as the driving force of evolution, his book *On the Origin of Species* sold out almost immediately and there had to be a number of re-print runs. Prior to this, although people were sceptical of the biblical stories of the origins of life there was no rigorous scientific theory that fully explained things. With the publication of his book in 1859, many people quickly saw the validity of his arguments, at last there was solid ground to build on. But to some extent that understanding was only on an intellectual level, the concepts of deep time and large changes, like a completely new species resulting from multiple tiny ones, was still difficult to accept at an emotional level because there was nothing to compare it to, they had no knowledge of biochemistry, genes or DNA.

Today we are in a more fortunate situation, we do know about these things and we are very familiar with modular design concepts. Take the computer for instance, we know at the lowest level that it is just electricity whizzing round circuits but we don't think about that when we are word processing or playing games on it. Computer chips are based on solid state transistors that react to electric currents in certain predictable ways. At the next layer up these are built into logic gates that provide certain outputs given particular combinations of inputs, at this stage just dealing with binary on/off patterns. The next layer up combines multiple logic gates into registers that can store and add together binary numbers. You could set up two binary numbers in two registers and press a button and their addition would appear in a third register; simple arithmetic has emerged from basic components, the whole being more than the sum of its parts.

This continues at each level up. With a few clever combinations, you can get subtraction, multiplication and division all from repeated additions and register inversions. Next, a string of instructions was put together and we had an assembler level program. Computer programmers used to write code in Assembler, actually telling the computer which registers to manipulate one by one, it was all very laborious. Assembler instructions are still there in modern computers (as are the chips and registers) but now programmers write code in high level languages and these are then translated by a compiler program into the basic assembler instructions. Programmers now only think about problems at a higher level, they don't think about registers and binary bits. Then, at the top level, we have someone using the program, say playing chess against the computer. At this level it almost seems as if the computer is a person playing against you, it looks like chess pieces moving around a chess board not programming instructions or binary arithmetic.

We are now very used to this modular approach, with layers of design and with the concept of emergence, the whole exhibiting features and abilities unpredictable by just looking at the sum of its parts. Computers have given us the mental concepts to fully take on board how evolution works. Cells for instance have many parts, a cell wall, a nucleus holding the DNA, mitochondria (the cell's energy storing bodies), mechanisms for constructing proteins, mechanisms for repairing DNA if it gets damaged and so on, the list is huge. Each part working by itself but together they make a cell. At the next level, different cells specialise into skin, liver, heart, bone, nerve etc. All the liver cells for instance together making an organ that again seems more than the sum of its parts, similarly with nerve cells making a brain. And at the next level up you have a complete animal with behaviour emerging that could hardly be predicted just by looking at its parts. *[For more on behaviour see chapter 10 on evolutionary psychology.]*

Like individual musical notes, given distinctive qualities, tones and harmonics by different instruments in an orchestra, brought together into phrases, melodies, counterpoints and harmonies and making up separate movements of say the Four Seasons by Vivaldi or creating the majesty of a symphony by Beethoven or the sublime architectural wonder of a Brandenburg concerto by Bach. The whole work in each case being a total emotional experience, so much more than the sum of its parts!

All of life's history on earth takes place in what is known as the biosphere. This is the atmosphere, the surface of the earth, the oceans and the top few miles of the earth's crust. It is an incredibly interdependent ecosystem, no part of which (except perhaps some bacteria) could exist by itself without the support of the other parts. That is especially true of the more complex animals, including us. We need food, we need oxygen in the air which is replaced by plants and we even need the bacteria in our digestive system. This biosphere of ours seems like a big place to us, covering the whole planet, yet if you represent the earth by something the size of a football, the biosphere would be just the thickness of a coat of paint on it!

From the first stirrings of life in the primeval soup of chemicals and amino acids on the early earth, emergent properties and behaviours have appeared hand in hand with the development of each new layer of complexity. At each level, these emergent properties could hardly be predicted just from looking at the constituent parts of the layer below but with billions of years to play with, natural selection has produced a fantastic variety of complex emergent behaviours and abilities within this cradle of life, this fragile, mutually dependent ecology of earth's biosphere.

Chapter 8

Where is ET?

In those days spirits were brave, the stakes were high, men were real men, women were real women and small furry creatures from Alpha Centauri were real small furry creatures from Alpha Centauri.

Douglas Adams

In chapter 7, Life's History, we looked at the history of life on earth. But is this unique, a one-off chance in a billion occurrence or is space teeming with life on other planets going round other suns? Maybe life itself is common but intelligent life rare, how can we know? It is extremely difficult to generalise from just one example, us here on earth.

There is an argument that says, look at the exponential scientific progress we are currently making and then extend this into the future. We will undoubtedly spread into the solar system and then find ways to reach the stars beyond. In fact, given just a few hundred thousand years at our current level of progress we should be able to colonise most of the galaxy. Now, if there is any other intelligent life in the galaxy, assuming some of them are older than our civilisation, they should have already done this. So where are they, why haven't they shown themselves? Since they haven't appeared, the argument goes, then they can't exist. Therefore it must be assumed that it is extremely difficult to evolve intelligent life. This is a reasonable argument and deserves serious consideration.

There is another school of thought that thinks aliens are already here. Throughout recent history, there have been strange sightings, alien abduction stories, crop circles and rumours of government cover-ups. Also, these days we are very familiar with the concept of aliens via science fiction books and films. People love a good story, especially if it involves a conspiracy theory, and there is no shortage of journalists, editors and authors willing to provide it without letting little things like truth and evidence get in the way. Even the US military was happy to spread rumours about UFOs in order to keep their latest aircraft designs secret. These stories, once started, seem to have a life of their own and run and run in spite of the copious evidence to the contrary. Even when some of the perpetrators of crop circles admitted they had made them, we still get articles about them every summer. Mystery sells!

Let's then get back to the more serious discussion – where is ET? As demonstrated in chapter 3, space is big, very big! We heard that there are more stars in the universe than there are grains of sand on all the beaches on earth. Given those huge numbers and the fact that sun is not very special in either type or location then the chance that life is unique to earth seems incredibly small. We are unlikely ever to know about other galaxies due to the distances involved but even our galaxy, given the billions of years it's been around, must have developed other intelligent life just from the sheer numbers of stars out there. So where are they? To my mind, there are two reasons we haven't seen them. One is a problem with timing and the other is to do with the difficulty of space travel.

The Timing Issue

It seems logical to say that since it has taken this amount of time since the universe began for us to evolve, obviously the rest of the universe is the same age so other parts which are conducive to life have had the same amount of time, so they should be at an equivalent stage to us. This reasoning shows the difficulty of comprehending the concept of deep time. We are only familiar with time as measured by our own experience, what seems like a lifetime to us is just a blink of the eye compared with just the age of the earth, let alone the age of the universe.

If we remember our paper scale for time from chapter 7, Homo sapiens have been on earth for just the last one fifth of 1 mm out of the 4.6 m since earth began. But the position of that fifth of a millimetre on the paper scale is very dependent on random events, chance and catastrophes. Take the meteor that killed the dinosaurs and allowed mammals to spread out into many more ecological niches, it could have arrived millions of years earlier or millions later or even not at all. Also, although traces of life appeared very early after earth was born, it took millions of years for simple cells to appear and almost half the age of the earth (2.2 billion years) before complex cells appeared. Many of the developments on the way involved random and fortuitous events which could easily have happened much earlier or much later by many millions of years. Our fifth of a millimetre could change position on the paper scale by a large portion of a metre either way, if indeed it ever gets there at all. This is an incredible difference in timing. On top of this, our earth was born 4.6 billion years ago yet the universe is around 13.6 billion years old so many of the other solar systems in the galaxy would have formed billions of years before or after ours.

So, to assume that an alien civilisation is at about the same stage of development as us, because they have had the same amount of time to evolve since the universe began, is completely to misunderstand the immense span that is deep time and the nature of contingency in evolution. Their fifth of a millimetre could be different from ours by a number of metres either way, a huge difference.

Given that, however, there is the question of how long their civilisation would last. This is difficult to estimate. From the evidence of life on earth, we get an average of five million years for a land-based species, that's 5 mm on our paper scale. Maybe an intelligent species would last longer because they can solve problems with technology, on the other hand they may have only a brief flowering of civilisation before their technology leads to wars or self-inflicted disasters like global warming? Then there are the random disasters like meteors or nearby supernova. It is a difficult question to answer but one would suspect that even an intelligent species will only have a short time span in respect to the incredible deep time of the universe. The only hope of persisting as a species for an appreciable length of time would be to spread out into space so that a disaster affecting one colony would not affect the others.

The above shows that we can't use the length of time argument to say that other alien civilisations, at a similar stage of development, should be around now just because we are. Indeed that actually works against them being around now. Suffice to say just that there has been more than enough time, by far, for intelligent life to evolve on other planets going round other stars in our galaxy. As we know though, species evolve and they also go extinct, likewise from our own history, civilisations rise and fall, so how many are actually there at this moment in the galaxy's history? A much better argument for other, current, alien civilisations in the galaxy is just by the sheer numbers of stars involved. But how can we tell what the probability is of other intelligent life in the universe?

Probability is usually expressed as one chance in so many, such as the probability of rolling a six with a dice is one in six. In more complicated cases probabilities can be multiplied together, for instance the chance of rolling a double six with two dice is one sixth times one sixth, which is one in 36.

Back in 1961, a scientist called Frank Drake came up with an equation for calculating the probability of other current intelligent alien civilisations in the galaxy. This is now known as the Drake equation. It involves assigning probabilities to different assumptions and multiplying them together to get a single answer. A lot of guesswork is required but it was assumed that as we found out more then we could refine our guesses more realistically. The equation involves finding probabilities for the following items: –

- The number of stars in our galaxy
- The fraction of stars that have planets
- The fraction of planets that can support life
- The fraction of those that life evolves on
- The fraction where life becomes intelligent after it starts
- The fraction of intelligent life that can and want to communicate
- The fraction of a planet's life during which a communicating civilisation lives

Multiply all these together and we get a number for the amount of communicating alien civilisations in our galaxy at any one time.

Let's try and put some numbers on these: –

The number of stars in our galaxy is around 250 billion, that is 250,000,000,000.

Many of these are small brown dwarfs (which probably don't have planets) or they are in binary or more combinations with other stars (which may well not support planets) or they just don't have planets. Because the light from planets is drowned out by the closeness of the stars they orbit, any planets outside our solar system are very hard to find. Our technology is only just beginning to find planets round the nearer stars so it is very difficult to generalise. Let's guess, however, that 20% of stars have planets, that is 0.2 as a decimal fraction.

How many can support life? In our solar system it is one out of nine, although we are looking hard for evidence of bacterial life on Mars and possibly on the icy moons of Jupiter and Saturn. Realistically though, a planet should be at the right distance from the star such that water can exist as a liquid on the surface. Also, it could well be that any planets orbiting stars too near the galactic centre are inhospitable to life due to the intense radiation there. Again we make a guess, this time say 1% of planets can support life, 0.01 as a decimal fraction.

Of those that can support life how many actually do? We saw from earth that almost as soon as it was possible, traces of life started. I would be very inclined to say if it can support life then it will. That is 100% or 1 as a decimal number.

How many of the planets with life go on to evolve intelligent life? Well we saw in chapter 6 on evolution that, although there is no real direction and no such thing as inbuilt progress to evolution, complexity does arise naturally, almost inevitably given deep time. Intelligent life, however, may well need a special environment, one for instance where natural climate variations is not too extreme, and meteor bombardment is at a minimum. Let's go for 1% or 0.01 as a decimal fraction.

What fraction of intelligent life can and wants to communicate?

Some life might be intelligent but live in the sea, like dolphins. Who knows? Let's go for 20% or 0.2 as a decimal fraction.

If an intelligent communicating civilisation does arise, what fraction of that planet's life does this intelligent civilisation last for? This is the really difficult one! First, say a planet lasts (like its sun probably will) about 10 billion years. But how long does such a species last? Well, we've seen on earth that land-based species average five million years but we have no real feel for an intelligent species. As well as natural disasters, they could bring on ones of their own, like war, global warming etc. Let's say the average is one million years. So one million out of ten billion gives 0.0001 as a decimal fraction. This is a real guess!!

Now, if we multiply all that together we get a figure of 100 current communicating intelligent alien civilisations in our galaxy. That sounds a lot but remember how big the galaxy is. One hundred in the galaxy means one for every 2.5 billion stars! So the chance of one being close to our sun is quite remote to say the least.

The Drake equation is really a bit of fun, you can put all sorts of guesses into it and you will get vastly different numbers out. Try it yourself, just set up a spreadsheet and maybe ask your friends what they think some of the guesses should be. Its real value lies not in the equation itself but in the questions it prompts us to ask when we try coming up with a realistic answer.

So, our best guess is that intelligent life is very likely to arise (purely due to the immense size of our galaxy) but that due to deep time, the number of such civilisations in the galaxy at the moment is very small. Even so, from the argument at the start of this chapter, given an accelerating technology (and assuming they don't destroy themselves) they should be spreading through the galaxy, so where are they? Here we come to my second reason for their non-appearance, the difficulties of space travel.

The Difficulties of Space Travel

Forget hyper-drive, warp factor, wormholes and faster than light ships, these are just devices used by science fiction writers to move their stories on at a reasonable pace. We can and probably will establish bases in other parts of our solar system within a few hundred years but moving out to the stars is a whole other ball game. Such journeys will take decades or even hundreds of years just to get to the nearest stars. You can immediately see that for any colonists this is a one-way trip and probably one that only their children or their children's children will see the end of (unless some form of deep freeze suspended animation is invented). We call such spaceships generation ships. Before you set out on such a voyage you want to be pretty sure there is a planet you can use at the other end, so first we would need to send out remote survey ships which can explore nearby stars and send back data. This remote surveying will also take many years but that is something we would probably do whether we send people there or not. This, however, needs a change in the way we think! We would have to send these survey ships out knowing full well that we will not get any information back until long after we are dead. We don't usually plan and do things on that timescale.

Then, any would-be colonists that follow would have to be completely self-sufficient in water, food and materials for hundreds of years and they would have to carry all the equipment and livestock etc to start over on the new planet. Now all this seems technically feasible, even if very protracted. The argument for spreading out into the whole galaxy in just a few hundred thousand years comes from the exponential nature of such exploring. Say earth sent out two colonies, then, after they were established for a while, they each sent out two colonies and so on. You only have to do this 20 times to get to one million planets colonised and allowing say 1,000 years between trips that is only 20,000 years. It soon builds up.

So, what is wrong with this picture? In a word, 'biology'. To make a go of a new planet the colonists would have to live there and grow food etc. We think of ourselves as autonomous independent units. We go anywhere on earth and live on what we find there, we visit other countries, eat the food and drink the water, with only the occasional problem like an upset tummy or a bout of illness. We treat ourselves as separate entities, almost machines, needing only nutrition and water to keep us going. It's not that simple, however, by a long way. We are part of this planet in a myriad of interconnected ways, integrally enmeshed in its ecology and ultimately dependent on its diversity. We co-evolved here, alongside all other plants, animals, insects, fungus and bacteria and we are family with them, sharing many if not most of our genes with each of them.

We ourselves are colonised by bacteria and other microscopic life, much of which we rely on. In fact, they make up about 10% of our body weight and they help us digest our food, providing vitamins and other nutrients we could not otherwise obtain. They also act as a line of defence, helping to stop harmful bacteria and viruses from attacking us. More than this, however, we are incredibly dependent on the whole ecology of bacteria, insects and minute animals that surround us. Insects which pollinate our plants, bacteria which help fix nitrogen in the soil, which the plants need, worms which create and aerate the soil and fungi which break down decaying vegetation. The plants themselves take in the carbon dioxide we exhale and give us back oxygen in return. We are uniquely evolved to tolerate the food we eat, digest it and make use of the particular mix of amino acids, fats, carbohydrates, vitamins, minerals and chemicals this planet provides via its complicated, interdependent ecology. Our planet contains millions of species of animals, plants and bacteria etc which provide an invaluable diversity of genes which we rely on.

Plants too have symbiotic associations with micro organisms, especially around their roots. Organisms which help plants extract nutrients from the soil and provide them with protection from disease. Also, some plants are co-evolved so closely with particular insects that only that insect can pollinate it and that plant is the only source of food which that insect relies on. If we go into space, how much of this can we take with us, indeed how much do we need? When we live in orbit, go to the moon or even Mars we take the food and water we need with us from earth and we can always get more supplies sent up. A one-way journey to the stars, however, is very different.

When we get to a new solar system and find a planet we might live on there are two possibilities, either there is no native life on the planet, not even any bacteria, or there is some sort of native life. Either way we will want to 'terraform' the planet to make it liveable on. On a planet with no native life, we would have to ask ourselves if that would be a good place to live in the first place. There would have to be a reason why life had not started, e.g. lack of liquid water, too far or too close to its sun, lack of atmosphere, too violent (volcanoes and meteors), poisonous atmosphere, too much radiation etc. In which case we might not want to live there either. But say we did find an empty planet that had the right conditions, firstly there would be no soil for plants to grow in and secondly there would be no oxygen in the atmosphere because there were no plants, bacteria etc providing it.

We could probably live in domes on the surface while we changed it but it would take thousands of years for any bacteria and micro organisms we released to start creating soil and oxygen. Managing that gradual release of organisms would be incredibly difficult. We would be trying to build an ecosystem from the bottom up at the same time as coping with them evolving by themselves in who knows what ways. Recreating an ecosystem anyway near resembling earth would be near impossible, the question is, would it be good enough to support the seeds and frozen embryos of the more complex earth life we had brought with us and eventually to support us?

Here on earth, every time we have altered an environment (either deliberately or by accident) through the introduction of a species from a different continent, it has almost invariably ended in disaster for a large part of that ecosystem. For instance, rats have often hitched a ride on ships and then escaped onto islands where they had not been before. There they wreak havoc by eating birds' eggs, plant seeds and small creatures, often causing the extinction of some native species altogether. An example of deliberate alteration comes from the Tahitian island of Moorea where a large snail called Achatina had been introduced as a delicacy but then escaped, multiplied and caused widespread damage to plants. They eventually decided to try controlling it by introducing another snail, the Euglandina, from Florida, which eats other snails. When released, however, Euglandina completely ignored Achatina and wiped out to extinction a native snail called Partula which lived on fungus and was harmless.

It is incredibly difficult to manage an ecosystem. If instead of a barren planet we found one that already had life, then it may well have soil already and oxygen in the atmosphere so we don't need the thousands of years to create these. However, we would have a whole different set of problems. That soil will contain its own forms of organisms and single-celled bacteria type life. This would very likely not be compatible with our plants and livestock. It would probably have some form of DNA based on amino acids but it would totally different to that from earth. It seems unlikely that our animals (or indeed us) would be able to make use of such native proteins and carbohydrates and without releasing bacteria we brought with us our plants would probably not grow. Also, what diseases and viruses are we likely to find? We may well have to start from scratch and we probably couldn't even live in domes on the surface to begin with because it would be impossible to keep these clear of contamination by the local flora and fauna.

So, the dream of colonising other planets in other solar systems is possibly just that, a dream. Or at the very least, it is incredibly complicated and of course, ET would face the same problems. But let's take a step back. To get to these other suns we need the generation ships. These would have to be totally self-sufficient, i.e. they couldn't take all the food they needed for such a trip, they would have to grow it on board. In fact, everything would have to be recycled: water, waste, air etc. They could afford to throw nothing away, it would have to be a totally closed system.

This in itself would be a complicated ecology. It would be easier to set up in the first place than a whole planet but being on such a small scale and with no back-up it could easily be ecologically unstable. They would not find this out, however, until it had been running for a number of years. The people would be carrying their bacteria inside them and the plants they grow would need their bacteria and micro organisms. Any animals would need the mix that suits them best, although if animals were carried as frozen embryos, some way of transporting their bacteria would have to be found as would ordinary soil bacteria, worms, insects etc. It will certainly not be possible to transport all varieties of such small but crucial species, so how do we choose. Also, on earth they keep each other in check. If we take a subset of micro fauna they could well become unstable as an ecology, indeed they would probably evolve and take on different functions from their role back home.

Another problem is that we need exposure to a large variety of these bugs in order to train our immune system. Without this, our immune system will not develop properly but more than this, it could also turn against our own bodies. We are already seeing a rise in auto-immune diseases in the West due to our ultra-hygienic lifestyle; things like asthma, diabetes, arthritis and multiple sclerosis. Our immune system is probably the most complicated part of our bodies and it is incredibly in tune with and intertwined with our environment. To put ourselves in an artificial or even just a highly restricted environment is to put ourselves at risk.

So not only is terraforming incredibly difficult, just getting there at all is extremely hard. We are learning more about genetic engineering and ecosystems all the time so our intrepid explorers may be able to catch the first signs of imbalance and put things right, even creating new micro organisms from scratch if required but we will need to know much much more than we do at present. Currently we seem hell bent on destroying earth's own ecosystems as fast as possible.

The best scenario I can think of for spreading into the galaxy is not to colonise planets at all. We would rely on space colonies. These would have to be large enough to carry a viable mix of organisms necessary to support life without becoming unstable plus large enough to carry enough people to make a viable society, including enough to keep producing more scientists and engineers as we go. This would be their home, not just a transport vehicle. It would have to be totally self-sufficient, until it stopped off to explore a new system every hundred years or so when it could stock up with water and minerals. This self-sufficiency includes factories for making tools and computers for instance. It would have to be incredibly hi-tech, including the ability to perform basic new research, if it didn't grow it could well stagnate. One option that springs to mind would be a captured asteroid that has been hollowed out and fitted with some form of propulsion yet to be developed. It would be spun on its axis to provide artificial gravity. NOT REAL!

As an extra precaution I see three of these flying together, each carrying a society of say half a million people to make it feasible. They would then be able to support each other in times of crisis. Each would probably develop its own culture and it may well feel quite exotic to take a visit from one to another. On reaching a solar system, they would explore it but they would also head for its asteroid belt to mine ores and get water from ice-rich ones. Here they would also begin construction of another three asteroid colonies and when the time comes to move on, two expeditions of three each would leave for two new destinations, thus slowly moving out into the galaxy.

The above is really a bit of science fiction, we are nowhere near capable of any of this and still it is incredibly difficult to manage even such a comparatively large ecology that has redundancy built in.

Another option for spreading into the galaxy is to do it remotely, by proxy. We currently send robotic rovers to Mars and spacecraft to other planets in our solar system. These are controlled from earth but they do have some limited intelligence built in. Any we send to other stars, however, will have to be completely automatic, taking decisions by themselves as to orbits, data gathering and what to point their cameras at. Even the nearest star to our sun is four light years away; if a spacecraft there asked us what to do next it would have to wait eight years for a reply!

Our computing technology is currently increasing in leaps and bounds. The best chess-playing computer can hold its own with the world chess champion. *[For more on artificial intelligence see chapter 12 on the future.]* If we do produce computers that are, to all intents and purpose, intelligent, then sending one on a spacecraft destined for another star would be much more efficient than sending people because you would not have the biology problem outlined above, and no need for complicated life support systems. It is still quite a technical challenge because you would want it to be able to repair itself if it went wrong. The easy way would be to send multiple spare parts with it but in the end this is self-limiting. Better would be to give it control of numerous robotic subunits that could mine ores from asteroids and manufacture new parts for itself. Indeed, if it could do this it could create an entire new copy of itself when it got to a solar system and then after exploring and sending data back to earth they could both set off in new directions. This is reproduction, it could almost be called alive!

Again this is science fiction but it does, on present trends, seem much more feasible. Indeed, it begs the question, not 'Where is ET' but 'Where are ET's AIs (artificial intelligences)?' Frankly, I'm a little surprised they haven't found us. Mind you, it would be like looking for a needle in a haystack for them. We as a species have only been around for a few thousand years, just a drop in the vast ocean of time. They would of course have to be programmed to respect life strongly wherever they found it or they could be dangerous even to their makers. Perhaps they are even now observing us from a distance!

Speculative ET

For another bit of fun, I'll finish this chapter with some speculation as to what ET might look like. Our biology is based on carbon. This element is fairly unique in the way it interacts with other elements creating an incredible variety of stable organic compounds that can themselves take part in other, more complex chemical reactions. There has been some speculation that one other element, silicon, could be used as a base for life albeit working at a different range of temperatures. Chemically, however, silicon doesn't produce as big a range of reactions. The likelihood is then that carbon will be the basis of any extra-terrestrial life.

In fact, as mentioned earlier, astronomers have detected organic compounds in dust clouds within our galaxy. These naturally form amino acids, which are the building blocks of proteins, indeed our DNA is protein built from amino acids. Where life develops on other planets, it will be because a protein, made of amino acids, has acquired the ability to replicate itself. As we saw in chapter 7, Life's History, life is built in layers with emergent properties appearing at each stage. At the basic level there would have to be something like our cell which would hold the genetic material inside a membrane and would organise chemical reactions such as energy production, replication, repair and transportation both ways across the membrane. A lot of these chemical cycles are very basic and would be the same as ours but then also many would be unique to the biology of that planet.

As multicelled organisms developed they would have to solve similar problems relating to more complex structures, as earth's organisms did, e.g. acquiring food, breaking it down, transporting it round the body, getting rid of waste, controlling and co-ordinating the whole activity. So, although the biological details within cells may differ from ours, the problems of complex organisms are similar and one would expect the emergent properties to be similar too. There would be a digestive system, senses, probably a nervous system, a skeletal system, an immune system, a circulation system, a respiratory system, a reproductive system and specialised organs to support all this.

Then we have the next layer up, a whole animal of some kind. On earth, we have many different animals, from fish to elephants and from insects to horses. The shape of these animals is not dictated by its biology, it is dictated by its environment, with a large degree of contingency thrown in. Take whales and dolphins for instance, their shape is very like a fish, dictated by their environment, although inside they are very much mammals, betraying their terrestrial origins. This is known as co-evolution, where two animals resemble each other (outwardly) because they have the same environment and similar needs. Some animals are specialists, like horses for instance. They live on the plains, eat grass and run to avoid attack. If something happened to disturb this environment, they may well not be able to adapt and could go extinct. Others are generalists, like rats, and would probably survive all sorts of catastrophes. One would expect an intelligent alien to be a generalist, as are we. We are not fast or strong but we have two things which make us excellent generalists – our large brains (used for anticipating trouble and finding food) and our hands (used for tool making and manipulating the environment).

One would expect similar senses – sight, hearing, smell, touch and taste, all-important to survival. The visible spectrum is where most of our sun's output of energy is concentrated and so an animal using this makes most use of light for the least effort, a standard evolutionary strategy. Some animals on earth can see in the ultraviolet and others can detect infrared so extensions at both ends of the spectrum are possible. The actual eyes could come in a variety of designs and may well be compound instead of camera type. Two eyes would be likely to give the binocular vision that is so useful. The senses would probably be grouped near the brain and probably high up so a head of some kind, with directional capabilities, seems inevitable. Other senses are possible, like some fish, which can detect electrical signals in its prey.

Gravity would play a big part. On a larger planet, you would expect the creatures to be shorter and stockier but on a smaller one, they could well be taller and lankier than us. This may also dictate the number of limbs. On a larger planet, ET could have four legs for extra support and two arms. The number of digits is optional and I leave that up to you. On earth, although the standard is five per limb, this could be an accident of history; there have been animals with six or seven and others with less. Also, there is the skin to consider, will it be smooth, hairy, furry or even some kind of scales or feathers. Again, the environmental history of the species will dictate, e.g. temperature, climate, habitat etc. There is a theory that humans lost their body hair and became smooth skinned due to a period in their history when they lived by the sea and spent a lot of time in the water. Skin colour will also be determined by climate.

We are looking for a species that has developed technology and the only way they could do this is with some sort of tool manipulation device i.e. hands of some sort, probably two but four is conceivable. The most obvious way for hands and hand-eye co-ordination to develop would be via a long evolutionary period spent in trees (or whatever the local equivalent is). Here it becomes a matter of life or death to be able to grasp branches quickly and accurately without falling. This is a common theme in evolution, where a mechanism evolves for one function (such as hands for grabbing branches accurately at speed) but then becomes used for other things (e.g. tool making).

Another requirement for a technical civilisation is that they are able to go through an industrial revolution of some kind. This implies that they can mine ores from beneath the ground, which in turn implies a physical shape which can dig and work in confined, enclosed spaces.

What about their behaviour? Again, I think they will be a lot like us. In all probability they will be gregarious for two reasons: one, if they work together they will achieve more but secondly and more interestingly, living in groups is probably a strong driver for the development of intelligence. You need to be able to keep track of a large number of individuals and your relationship with each, whether you are in their debt or they in yours etc. Gossip about other members of the group needs to be remembered and used at the correct time. Intelligence is also required to predict what others in the group will do in given situations, an ability to see things from the other person's perspective and an ability to plan things, like hunting food.

As we see on earth, a huge variety of animal shapes are possible but if we want to find an intelligent, communicating species then only certain physical attributes will fit the bill. They are likely to walk upright on two legs, have two or four arms with some form of multidigit hands and have a head with sense organs. They may or may not have a tail. We can't say much about skin colour or skin covering. In short, they will be different from us internally, but there are likely to be many external similarities not only in looks but also in behaviour. They will be survivors, which will give them an aggressive side to their nature (like us) but they are also likely to be gregarious and co-operative with a strong sense of empathy and an innate curiosity about life. Again, this is a fun thing to speculate about and I'm sure you can come up with a few ideas of your own, maybe a new sense or something. Culturally they will probably vary widely, just as we do across the globe. And what art, music and poetry will they have? Fascinating!

Summary

Why have I included this chapter in this book? Currently we know of only one example of life in the universe and that is life on earth. One of the aims of this book is to seek to understand our origins. So is life special to earth, are we special because we are the only intelligent beings in the universe? In the past, it was thought that earth was special, that it was the centre of the universe. Now we know it is just a small chunk of rock going round an ordinary star two thirds of the way out from the centre of one galaxy among hundreds of billions of galaxies. Scientists are looking hard to find other life. There is a large program going on searching for radio signals coming from space and the probes we send to other planets and moons in our solar system are also searching for evidence of basic life.

Evolution is a general organising principle; if life does start elsewhere in the universe then evolution and natural selection will get to work on it. It is early days in our search and the galaxy is huge; it's like looking for a needle in a haystack without actually going to the haystack. However, it is that very hugeness that virtually guarantees that life and intelligent life does exist out there. Go back to our guess with the Drake equation; we calculated that at any one time there could be about 100 intelligent species in our galaxy, not a lot in such a big galaxy. Now remember, however, that there are around 100 billion galaxies in the universe. Just look up into that night sky – there could easily be, at a conservative estimate, trillions of intelligent species looking back at you and asking the same questions! Are we special, chosen even? I don't think so!

Chapter 9

Culture

Whenever I hear the word culture… I release the safety-catch of my Browning!

Hanns Johst (German playwright 1890 – 1978)

Culture is a huge area of study, incorporating as it does a rich mixture of art, music, literature, social sciences, politics, traditions, customs, behavioural mores and much else besides. In this chapter, however, I want to cover only one aspect of the subject, and that is the general method by which culture originates, spreads and evolves over time. We learnt in previous chapters that we have hardly changed as a species since we became Homo sapiens in Africa around 200,000 years ago. There has probably been some small genetic drift and some extra genetic variation added to our gene pool since then but essentially we are still the same animal. However, as a society we do have the very strong impression that we have become more intelligent over that time and indeed over just the past few hundred years.

I demonstrated in chapter six how the concept of evolution has become unfortunately tied up with the Victorian idea of progress. Look at how civilised we are, look at our technical achievements and compare this to savages in the jungles, the argument goes. Actually, those natives are no different from us, they have their own accumulation of knowledge and we would be as hard pressed to survive in their environment as they in ours. The difference is one of culture.

In chapter two, we met the concept of memes. Whether there are such things as memes is a matter for debate and indeed the debate rages, with many people picking up on the concept and expanding on it. Either way the word is very useful as a shorthand term meaning a 'unit of culture' and I will continue to use it here. As mentioned before a meme could be an idea, a belief, a fashion, a way of behaving, a set of instructions for making something, a tune, in fact anything that is basically an idea or set of related ideas that can be passed from one brain to another. Cultural evolution can be compared and contrasted with biological evolution, with memes being somewhat analogous to genes. There are many similarities but there are also a number of instructive differences.

For instance, whereas biological evolution is strictly Darwinian, cultural development is very much Lamarckian, that is memes don't have to wait for random mutations, they can be added to, changed and expanded upon by each and every mind they come to reside in. Like genes, any changes are not necessarily in the direction of 'improvement' in any intrinsic sense. If a change means that a meme will spread better then that meme will survive for longer, whatever the change was. As we saw in chapter two, memes can work together in a memeplex and if the memeplex includes specific memes that protect that memeplex (like 'thou shalt not question the word' and 'it is a virtue to have faith') then the memeplex will probably survive virtually unchanged for many years.

Being Lamarckian in nature, however, even religious memeplexes do change, splits occur, different sects emerge and interpretations change. As people move round the world there is a cross-fertilisation of ideas with the new cultures they join. Good ideas spread very quickly and, with many people becoming involved, they are improved upon quickly too. This process gets quicker and quicker in an exponential fashion and it is this development of culture and society that gives the impression that people are becoming more intelligent.

Trade also helps the spread of different cultural ideas and has done so right from the earliest days of human society. For instance each tribe would probably make their own tools, and you have only to see one person with a stone axe to get an idea of how to make your own. But then you may well come up with improvements to the method, which you would probably share with your fellow tribe members. It is entirely possible that one tribe would get the method down to a fine art and their axes would become renowned in the region as being strong and reliable. Although other tribes could make their own axes, it may well benefit them to concentrate on something they are good at and trade that for the axes from the first tribe. Here we have the start of more complex societies with division of labour, trade and specialisation.

Then came agriculture. This started when people realised that they didn't need to be nomadic, they were able to keep their animals in one place and grow their own food. Farmers soon discovered better ways of doing things and improved on their methods; new varieties of wheat were developed and more productive animals were bred for. Ideas for doing things, memes, are constantly being improved on in a cumulative Lamarckian manner. A similar process occurred with the Industrial Revolution. All the time memes are being passed around, modified and improved upon but in no way are we becoming more intelligent; biological evolution just doesn't happen on anything like that timescale, this 'improvement' is all down to cultural evolution.

If we remember, for movement in a particular direction biological evolution requires a selection pressure, which in turn means that those of the species that can't keep up will be left to die out, which just isn't happening, people aren't currently dying (or not breeding) because they are not intelligent, therefore there is no current selection pressure for growth in intelligence within our species. In fact, almost the opposite could be occurring due to the fact that more people are surviving now where in the past they may not have done. This would not just affect intelligence but other features of human beings as well, like for instance our immune system. *[See chapter 12 on the future for more on this.]*

A particularly interesting aspect of culture is the tendency for the development of groups. The 'us and them' phenomenon. People naturally seem to join together in like-minded groups, for instance: clubs, communities, football supporters, nations, religions etc. The basis for this (as in most behaviours) is a mixture of both genetic and cultural (i.e. environmental) influences. There are great benefits to be gained from working together in common cause and genes discovered this early in evolution *[I'll explore this genetic aspect of behaviour in a lot more detail in the next chapter.]* It is extremely difficult to separate out the genetic and cultural drivers for any particular behaviour, however, here I only want to talk about the cultural influences (while at the same time remembering that genetic ones remain in action alongside them).

Continuing the idea of memes working together in memeplexes, we see that many group-orientated memeplexes include memes that protect that memeplex, helping it to spread. For instance, if you join the group there will be benefits for you, and others in the group will treat you well (certainly a lot better than they treat outsiders). If you leave the group you will be treated very badly, a deserter or betrayer. Religions have a word for some who leave that religion, they are called apostate and in some cases this even brings a death sentence!

Co-operation can be very good for the group and hence for the memeplex, helping both of them prosper and spread. However, this has also been the cause of much misery in the world because of the 'us and them' nature of such memeplexes. A very strong fear of strangers and outsiders can easily develop. In nationalism, for example, you get memes like 'immigrants come in and take our jobs' and in religion there are memes like 'non-believers will contaminate you' and 'our religion is the only right one'. Both nationalism and religion have been, and currently still are, large causes of conflict in the world.

Our current scientific period is one of even quicker change, mainly because (unlike religion) science has the meme of 'constantly question all assumptions' built into its memeplex *[See chapter 11 on the scientific method.]* Science continually strives to improve itself, it's a fundamental aspect of its philosophy that it should try actively to do so. Western society is now so specialised in its division of labour that it would quickly break down if some catastrophe occurred. For instance, we are all very dependent on a few multinational companies that make computer components, TVs, phones and cars etc. If we lost the capacity to make computer chips it would affect the whole way we live our lives; our daily business is increasingly dependent not just on the computers on our desks but on the Internet as well. You might be able to make a stone axe yourself but a computer chip is another thing completely, even if you knew in principle how it worked it requires a whole industry to support its manufacture.

It is the scientific memeplex that has given us our current understanding of the universe and our place in it. Two elements here intertwine and feed off each other. One is our scientific tools, which have become successively more powerful and more accurate over the past few hundred years, starting with the telescope and progressing to particle accelerators, electron microscopes, DNA sequencers and computers. People just keep coming up with more and more ideas as to how to improve the tools or build better ones, all incredibly Lamarckian. Hand in hand with this has been the successive refinement of our scientific theories as we interpret the data our tools provide us with. Again, many people getting involved and suggesting improvements.

These cultural advances in agriculture and industry together with specialisation and the division of labour have, as a by-product, given us more leisure time, which in turn has allowed a flourishing of other aspects of our culture. Music, art and literature have all developed tremendously. These have of course been enriched by geniuses such as Mozart, Turner and Shakespeare but they too had their influences from previous generations.

Science too has its share of geniuses, like Newton and Einstein. In fact, Newton once said that if he could see further it was only because he was standing on the shoulders of giants who had gone before him. When Einstein came up with his theory of relativity there were only a handful of mathematicians in the world who could understand it. Now it is taught to students at university and many theorists round the world are working on extending it and combining it with quantum mechanics. Differential calculus, developed by Newton and Leibniz, is taught these days even to children at school. It takes genius to come up with something new and revolutionary but once the meme is out there, others can understand it, build on it and even simplify it. Thus is the illusion of increasing intelligence. In reality, it is the Lamarckian nature of culture that distorts our perception providing a false perspective that is both comforting and self- flattering. This is not to say we aren't intelligent, we are, and it is a tremendous and fortuitous gift from evolution, only that we need to use it logically and wisely as we take on board what science is now finding out about the true nature of the universe and our place within it.

Chapter 10

Evolutionary Psychology

We are built as gene machines and cultured as meme machines, but we have the power to turn against our creators. We, alone on earth, can rebel against the tyranny of the selfish replicators.

Richard Dawkins

This book's twin themes of 'Where did we come from?' and 'What is the reality behind our perception of the world?' are both brought to light in this chapter, because how we think and how we perceive are intimately tied up with and indeed designed by our evolutionary history on earth.

A second reason for including this chapter is to provide an insight into how the brain works, both from a processing perspective and a behavioural one. Without this understanding, the brain is just a magic black box and it would be tempting to attribute supernatural explanations to it, such as a controlling spirit of some kind. As is often the case, fact is more interesting than anything we could make up and the details of how the brain functions and where behaviour comes from are fascinating. I don't want to give the impression that brain science is complete, we are still learning more all the time but we do know far more than we did just 100 years ago. Again, this is mainly down to the incredible improvement of our tools.

(A word of warning here. The phrase 'evolutionary psychology' has been used in many ways and in its strongest definition it is employed to suggest that everything we think and do is a consequence of our evolutionary history, especially that part of our past where our ancestors dwelt in the plains of Africa. This definition ignores the nurture side of the nature/nurture debate. We are shaped by our upbringing and the random nature of our personal biological and psychological development as much as we are by our genes and instincts. My use of the phrase is in a more general sense, concerning how a brain functions and how this functioning has developed through evolution and natural selection.)

It is often our impression that we are logical, rational beings, that we control our environment and make decisions based on free will and on an analysis of the situation. What type of decisions do we make? They are often about things that we perceive to be in our best interest or to satisfy our needs and desires. Most of those needs and desires, however, arise subconsciously, in fact a large part of the processing in our brains goes on below the level of our conscious mind. Even something as seemingly innocuous as deciding what film to go and see has much going on in the background, things like why go in the first place, who to go with, friends, family, bonding, courtship and a myriad of other factors.

Our subconscious brain deals with many things without our awareness of them, which is a good thing otherwise we would be swamped with trivial decisions to make all the time. It caters for the day-to-day, second-to-second even, running of our bodies only bringing things to our conscious attention when required, like needing the toilet or becoming hungry. The subconscious brain is the origin of our desires and our instincts, it also automatically handles a huge amount of the pre-processing of the sensory information coming in about the world. We have the distinct illusion that we see the world as it really is, almost as if our conscious mind is sitting somewhere in our head watching a video screen of the outside world and making rational decisions as to what we should do. Obviously it's not like that, if there was a little man in our head doing that there would also have to be an even smaller one in his head doing the same and so on ad infinitum.

So, what are we and why do we think and act in the way that we do? Basically we are animals with a psychology designed by natural selection based on the need for survival and reproduction. As a by-product of these two needs, however, and enhanced by chance developments within our evolutionary history, our psychology includes a number of subsidiary aspects, for example: –

- A capacity for learning
- Curiosity
- A capacity for awe
- An appreciation of beauty
- A capacity for logical thinking
- A capacity for love

These subsidiary aspects arise directly from our two prime needs and were developed specifically (by natural selection) to enhance our survival and reproductive prospects. They did not evolve because they were good things in themselves to possess. Therefore, when considering these subsidiary aspects of our psychology, we should keep in mind that they are all based on serving either our survival or our reproduction. This is not to denigrate these aspects of human psychology which, having arisen, are very real and valued attributes, it is simply that it is important to understand why we think about things the way we do. It should add insight and an extra dimension to our thoughts as well as adding an appreciation of their limits.

In looking at the brain I'll start with an overview of neural processing in general, relevant across all areas of the brain, then I'll look at the modularisation of the brain and how this is co-ordinated, and finally I'll look at the origins of behaviour.

The brain has many separate components that each does a specific job depending on its structure and the way it is connected but in general and across all these specific components the way our brains work can be broken down into three main processing concepts: pattern recognition, model building and the processing of symbols. These are layered, with each type of processing building on the previous one.

Pattern Recognition

A capacity for learning is the most important aspect of a brain and was one of the first parts to evolve. If you can't recognise danger you won't survive long and if you can't identify a mate you will leave no descendants. Therefore, because we are here, all our ancestors throughout evolution must have been good at both things. To be able to learn, two things are necessary. First, you have to remember past events and secondly you have to compare your memory to what's happening now. In other words, you need to be able to recognise patterns. If one thing followed another before, it is very possible it will again, for instance if you see a lion you had better very quickly learn to run. Being able to identify hidden predators has tremendous survival value.

Pattern recognition is fundamental to the way the brain works. Although the brain has many different components, and separate areas of it are used for processing different things (like vision, sound, smell, touch, taste, balance, movement and planning to name but a few), all of these areas use pattern recognition as their main processing tool. The brains of even the smallest animals all work in a similar fashion using pattern recognition as their main tool.

How does the brain perform this pattern recognition? Again, here computers have helped our understanding of how our brains work. A new development in computing is called neural processing. This is an attempt to model the way neurons in the brain connect to each other. Each neuron in our brain can connect to as many as 10,000 other neurons and when it is stimulated, it in turn stimulates many others in a kind of cascade effect. Some paths through this cascade are stronger than others, a bit like a forest where you can go any way you want through the trees but some paths get well used and become more permanent. In the brain, if an outside event stimulates one of these more permanent paths, this will probably be because a similar event stimulated the same path before, we experience the recognition of a pattern.

In the human brain there are around 100 billion neurons and each one has between 1,000 and 10,000 connections to other neurons, mostly to neighbouring ones but some to more distant ones as well. In computing we have been able to model this (on a much smaller scale of course) creating a neural network of connections and giving variable weightings to different paths. We have then been able to train the network to recognise a picture of say an individual face by showing it many faces and telling it which one was the correct one each time. The program would adjust its own weightings on its neural paths and would learn which weightings and paths in the network corresponded to the correct face. After training, the network could, more often than not, pick out the correct face from a selection, even if the picture was taken from different angles.

So here we have pattern recognition performed by a simple neural network without the normal linear logic computers usually use. So far it is at a similar scale to that of an insect's brain; the human brain is still many times more complex than our best computers. However, as we saw before, much of the brain's processing is at a subconscious level and this is carried out by many different subsections in the brain, each dealing with its own area of speciality. Each subsection comprises only a small number of neurons compared to the brain as a whole but as seen with the computer neural network you don't need that many neurons in a network to get simple pattern recognition. The computer example dealt with vision but the principle is the same with any event that happens: hearing, smell, touch, taste or something internal like emotions or logical reasoning. When we recognise patterns one of the things we are attuned to is cause and effect, again being able to identify this has tremendous survival value because we can then anticipate an event before it occurs.

Having developed this pattern recognition ability for survival reasons, we can now make use of it for anything else we choose. These days we deliberately train our brains in things we deem important, for instance by sending kids to school. Think of learning our times tables for example, not something we would come across in nature but with training we soon burn in the correct paths. It should be noted that pattern recognition errs on the side of caution and does so for good survival reasons. Even if something might be a pattern, it gets flagged up for attention in case it is important. Is that a tiger moving in the undergrowth or just a pattern made by the reeds waving in the breeze? We are always looking for patterns and often see ones that are not there, for instance invoking supernatural explanations for things that happen, like a bad harvest; perhaps we didn't perform the correct sacrifices to the gods! This is where science triumphs, with its insistence of performing repeatable experiments under controlled conditions and its constant testing and re-testing of any theories that are generated. *[See chapter 11 on the scientific method.]*

In the example above, the computer neural network had to be trained before it could recognise the face, it had to first be told what face was important from a number of examples. So the question arises, how does an animal know what is important to remember out of all the things that happen to it every minute of every day? How does the brain set its own weightings, decide what paths should be burned more deeply? The answer here is to do with emotions. There is a structure in the brain called the Amygdala which is central to our emotions. It has neural links throughout the brain and it can stimulate the release of chemicals related to our emotions such as adrenalin (the release of which makes us suddenly more aware and primes us for the fight or flight response).

When we are subject to a strong emotional response to something, any current mental states or patterns are stored more strongly than when we are emotionally neutral. That particular event means something to us, it is important to remember it. And, when recalling such a memory or when seeing/experiencing the same thing again, the emotion is also recalled. This is a strong mutual feedback process that acts to prime and reinforce important memories.

In fact, our brains are awash with a complex cocktail of chemicals and these are all extremely important in laying down memories as well as affecting our moods. Some of these are the so-called neurotransmitters, they are found in the brain at the junctions of nerves. Nerves don't actually touch each other. What passes the electrical signals across the minute gaps between them are the neurotransmitter chemicals. The three main ones are called serotonin, dopamine and norepinephrine. These can be present at different concentrations and they strongly affect your mood and how you remember things. Serotonin affects whether you are feeling confident or defensive, dopamine regulates things like your tendency for novelty-seeking activity and norepinephrine is involved with our reward and craving subsystems.

Other neurochemicals, like the fight or flight hormone adrenaline we have already met, include oxytocin which is involved in the forming of emotional attachments, endorphins which are the brain's own painkillers, cortisol which is part of our stress response subsystem and of course the more familiar hormones like testosterone and oestrogen. These are just some of the chemicals in our brains, all of which regulate and colour our moods, the way we interpret our experiences and how we lay down our memories. The very memories we rely on for pattern recognition.

Model Building

As we consider more complex animals with larger brains, we find a step up from simple pattern recognition. Patterns and memories get linked together and the animal starts to build models of the world it finds itself in. For example, we learn to use the input from two eyes to estimate distance and three-dimensional depth in what would otherwise be two flat pictures of the same scene from very slightly different angles. Having created that model of depth in the world, we can automatically mentally compensate for the fact that a familiar object in the distance is smaller than when it's closer. That speck in the distance really is a lion and is just as potentially dangerous as when it's closer and looks bigger. We have a model of the world that includes things like lions roaming around and we can automatically estimate its distance and the amount of time we might have to evade it.

We build models of all sorts of things: our field of view, our house, a map/model of our neighbourhood or even a model of a friend; how she thinks, behaves, what she is likely to say about something. These models seem to build themselves as we recognise, refine and re-refine more and more patterns and memories.

I remember how I came to know London. At a time when I didn't live there, I would visit occasionally, to go to a show or a museum or something. Initially I would go to the place I wanted on the tube trains and gradually I built up models of the streets around each tube station. Then one day I wandered further afield (should that be atown?) and suddenly recognised the street I was in. It was part of another model of the streets around another tube station. I knew where I was and there was a kind of an 'Ah ha' moment, so that was how the two models fitted together. There was a sort of re-jigging of the two models in my brain to create a single, more accurate model. Later, when I actually lived in London for a while and walked and cycled more, this happened again and again as the isolated islands of streets, connected by linear underground lines, gradually re-orientated themselves into a coherent model of the city. I could then think about this model and plan a route through it at will.

Note, however, that this model building, like pattern recognition, is a largely unconscious process. It's going on all the time as we unconsciously become familiar with places or people and constantly update and re-organise our memories of them. Memories have to be both reliable and pliable at the same time, an incredibly fine balancing act that is almost totally unconscious. It is of course an evolutionary adaptation that confers great survival value. Cognitive scientists have demonstrated in numerous examples that this continual updating of memory shows just how unreliable witness statements can be with constant recalling and probable update; also, we love telling stories so such elaboration is almost inevitable. In fact, it has been found that therapists can even create memories that weren't there before. This is a basic part of our model building of the world, memories have evolved to be both dependable and updateable!

Another analogy here of how we build models is the idea of hooks. This is where we attach new memories to existing ones, like hanging them on a hook, thus making for easier memory storage and retrieval. For example if you wanted to just remember a string of numbers it would be quite difficult unless you had something to 'attach' it to, but if it was the phone number of a person you know it becomes much easier, it gets added to your collection of facts about that person, your internal model for them. It may well become easier still if it links into other models, for instance if they live in the same area as another friend the area code of the number will be the same.

Symbol Handling

Most animals build models of their world, some more sophisticated than others. Where humans (and some other primates to a lesser degree) have triumphed is in taking this to the next level up, i.e. symbol handling. This is where a whole concept is represented by a simple idea or word. Humans are uniquely good at this largely due to the development of language. All brains can recognise patterns, many can build internal models of their environment but few can handle symbols. A dog for instance can handle symbols to a small degree; it will understand when you say the word 'walk', it may even go off and fetch its lead, but it is humans that have made an art form of symbol handling.

The obvious case of this is from algebra, where we use the symbol 'x' as the unknown. Even though we don't know what x represents, we can manipulate it. We can add to it as in x+1 or we can talk about more than one of them as in 5x. We can include it in complicated equations and pass the results to each other, even though x remains undefined. The same is true for the numbers themselves. If we say three apples, everyone knows what we are talking about, there is a concept of 'threeness' that we have learnt as a child. Even though three apples is different from three chairs we are confident to use the symbol three to convey information to another person in an incredibly useful manner.

In fact, every word we use is a symbol for some concept or other. You can say 'she sat on a chair' without having to define the word 'chair' (or even 'she', 'sat', 'on' or 'a'). We manipulate symbols as complete entities in their own right to convey important information to each other. Someone might say to his friend, 'I would not normally vote Tory but I will this time because the candidate is a Euro-Sceptic.' Now 'Tory' and 'Euro-Sceptic' are complex concepts compared to 'chair' but at a high level, everyone knows what the speaker is talking about. Once a symbol is defined, even in a loose or fuzzy manner, it can be used (just like 'x') in a sentence and it can be manipulated.

Language is very fluid, however. If you get into a conversation with the speaker in the previous paragraph and dig down into his use of these words, you may well find nuances that are different from your expectations. Then between the two of you a clearer understanding of the definitions of the words will arise. These then become the definitions you understand between the two of you when you speak of these things but with someone else, you might have a slightly different understanding when these words are used. This is true not only for complicated concepts but for seemingly simple ones as well. Take the word 'chair' for instance, how would you define it? A piece of furniture that is used for sitting on. It must have four legs not three or it would be a stool, but wait a minute it could have wheels. Is it wooden or padded, what's the difference between a chair and a throne? Are we talking about a theoretical chair as in chairing a meeting? Context is everything. If two people need further to define the word between themselves then they will do so and, in that context between the two of them that will be the meaning. The important thing here is the conveyance of information, as long as that is achieved then the word has done its job. There is no such thing as 'chair', it is just a concept and a fluid one at that, as are all the other words/symbols.

It is an important distinction to make that a word is not the thing it represents, it is just a symbol used to convey meaningful information. You might understand the word 'wet' but that is very different from actually putting your hand in water and experiencing wetness. We get so carried away by words that we sometimes forget to experience the world directly. Words are really powerful but it is also easy to pick up the wrong meaning when a person says something (as they say, diplomacy is an art). Language is changing all the time, the words themselves are pliable and new words come into common usage. A language is almost a living entity but it is all just symbol handling using fuzzy, shorthand concepts to convey information from one person to another and we humans are past masters at it.

Another symbol we use is money. Those notes and coins in your pocket are really just bits of paper and lumps of metal that we agree between us to have a certain value. They are not valuable in their own right, you can't eat them, they are only useful because other people agree they are. This is backed up by the laws and punishments society has put in place to ensure these things are not taken advantage of. Money is a symbol that we all understand and use and how much you have is an indication of the value that is placed on you by society (note there is no ethical judgment implied here).

It is the ability to handle symbols that has enabled us to create our complex culture. We can throw concepts around and pass information to each other in a way that no other animal can. This ability and the development of language must have been a powerful driving force in our evolutionary history, a bit like prey running faster to out-pace the predator running faster. What started as chance evolutionary changes to vocal cords, facial shape and facial muscles must have given early hominoids great survival advantages, especially in their ability to work together as a team. And this in turn could have driven an enlargement of the brain to cater for such logic and symbol handling. Symbol handling allows us to talk about things at a high level without being bothered by details.

From simple pattern recognition, through model building to complex symbol handling, our brains, largely unconsciously, have developed a tremendous capacity for learning about the world and an innate natural curiosity, both of which arose in the first place for their survival value but which, having arisen, we can now use in whatever way we desire. This is a common occurrence in evolution, where a function or biological structure evolved for one purpose gets co-opted for a place in a new function.

We have looked first at the general processing that brains use across all their areas. Another aspect of brains, however, is how they subdivide their work into separate areas or modules. This can most clearly be seen with the visual processing system.

Vision

From an evolutionary perspective, sight is extremely important to an organism. Animals need to find food, find a mate, avoid predators and avoid other dangers in the environment like cliff edges. Also, if you live in trees as monkeys do for instance, you need extremely good three-dimensional vision and depth perspective in order to jump from one branch to another. For these reasons a very large area of the brain (towards the back of your head on both sides) is given over to visual processing. The operative word here is 'processing'. We have the distinct impression that what we see is an exact picture of the world as it comes in through our window-like eyes but this is an illusion, it doesn't work like that. A tremendous amount of pre-processing occurs to this incoming visual information, all at a totally subconscious level, well before we become consciously aware of what we are seeing.

For a start, it is not a 'picture' that comes into our brain at all, it is just electrical signals travelling along nerves from different parts of the eye's retina. Also, with two eyes, each giving a slightly different view, two sets of incoming signals need to be combined into a single image. In fact, what actually happens is that the incoming information is split up and routed simultaneously to quite a number of very specific areas of the brain, each specialised for a particular task.

There are special brain areas for colour processing, movement detection, depth perception, shading, edge identification (usually an indication of the boundary of an object), line orientation, object tracking, picture stabilisation (to compensate for head movements, as when walking) and many other functions. One of the beauties of this modularisation, apart from the ability to specialise, is the power of multiprocessing. All the modules are constantly processing the same incoming information at the same time, rather than one after the other like a computer. This is what gives us our virtually real time, on the move, constantly updated visual picture of the world.

What we end up 'seeing' is not an image but a real time model of the world, built not just from incoming data but also from learnt experiences as to how that data should be interpreted. For instance, a dark circle that is getting rapidly bigger in front of you is not an object that is changing size but a ball that someone has thrown at you. This is the model building that we met earlier. A baby would not know about a ball or about depth and distance or that something far away only looks smaller, all this has to be learned through experience. The baby would just see a growing circle in its field of vision. All these things come together in the working model of what we 'see'. If for instance you were colour blind you would probably have some defect in the colour processing module (many things could cause this) but you would still 'see' a coherent working model of the world.

Try an experiment. Put two black dots on a piece of paper about an inch apart. Close your left eye and slowly bring the paper towards your right eye concentrating on the left-hand dot at the centre of your vision. When it gets to a particular distance the right-hand dot will disappear! All you will see is white paper where the dot should be. The reason for this is that the right-hand dot is being projected onto the back of your eye at a place where there are no receptor nerves and that is because this is the place where all the eye's nerves converge into the optic nerve to pass through the back of your eye into the brain. Now, the interesting thing here is not so much that you don't see the dot but that you do see white paper, no hole, no blackness, but 'invented' white paper! The brain has created this as part of its visual model. Indeed everything that you are 'seeing' is in fact a model of the world in your head, all be it a pretty good one or you would not survive.

For an experiment once, a researcher created a special pair of glasses that turned the world upside down. He wore them constantly for a number of weeks and very soon, he was 'perceiving' the world the right way up. He even learnt to fly a light aircraft while wearing them. Then when he finally took them off he saw the world through his own eyes as upside down until he finally got used to it again.

All this parallel processing takes place at a subconscious level, which of course is necessary not only because it has to turn electrical signals into a viable model of the world but also because there would just be too much data to be conscious of all at the same time. We would be overwhelmed and not able to function (i.e. survive) so evolution has given us these mechanisms for handling all that data. In fact, we can only be consciously aware of four or five things at any given time, our brain just fills in the rest (like the dot experiment) as if the continuous picture of reality is there all the time.

An interesting experiment was done recently in America where a number of volunteers were asked to watch a video of a basketball match. They were split into two groups, one group was just told to watch the match while the other group had to do something specific, like count the number of times one side touched the ball. At one point in the match, someone dressed in a gorilla outfit came onto the court, danced around for a bit and then left. The group just watching the match all recalled this when questioned afterwards but of the group who had to count specific things, around half had no recollection of the gorilla incident and almost refused to believe it had occurred at all.

So how is all this co-ordinated? Rather than a single controlling entity sitting in your head it is much more like a parliament of modules coming to a group decision as to what is important or what our attention, decision making, memory and planning subroutines should be focused on; each module speaking up for its specialist area. For instance the 'movement detecting' visual brain area will most of the time be fairly quiet but if something happens at the corner of your vision it will suddenly shout louder and you may well turn your head for a quick check: is it a tiger or just some waving reeds?

So it is worth re-stating that what you 'see' is not a 'picture' of the real world, it is a 3-D model that is made up inside your head which is being constantly updated (subconsciously) with electrical signals from your eyes, electrical signals from your balance system and a lot of pre-processing involving memory and learned experience about the world (for example perspective correction). This model is not even the real picture, it is just what our senses are tuned to. For instance, many insects can see ultraviolet light, we can't but it is there all around us. The insects see the ultraviolet colours of flowers but we can't. We also don't see infrared or x-rays or many other wavelengths. What we think of as our 'real' view of the world is just that information that natural selection has decided is the best we need to survive in the world at the least cost in processing terms. You are looking at a model!

It is interesting to speculate how bats for example perceive the world. They use echo location to find or avoid things in the dark.

Their incredibly complicated audio processing will for them be the major input to their 3-D model of the world. It locates not only distance but also the texture of surfaces, perhaps a bit like our colour perception. I would bet that their echo location to them is very like 'seeing' is to us.

This subsystem processing approach is not just reserved for vision, it applies across the brain. Information coming in from other senses gets split up and subconsciously processed in the same way and built into our constantly updated model of the world. Touch is a good example. These areas have been extremely well mapped out in the brain with each area of the body being wired to very specific brain locations. This wiring occurs during the development of the foetus so these specific locations are the same in all of us. But after this initial 'wiring up', when we start to use our brains things become more individual, well-used paths get burned in. Think of a violin player training her movements and her understanding of music; the basic wiring is there to start with but specific paths will be developed that someone else won't have.

Going back to touch and consciousness think about that chair you are sitting on; you can feel it with your sense of touch when you think about it but most of the time you are not thinking of it even though the information is still coming in. If it started to heat up though, it would slowly seep into your consciousness until you realised that it had been getting hot for a while and that you had better do something about it.

So, it's not just your vision but it's your whole brain (including your memory) that works like a parliament of modules all calling for your attention and your focus to a larger or lesser degree. Daniel Dennett gives a good example here: you have been hiking all day and have a blister on your heel. You are aware of the pain and of your fatigue. Then a mother bear and her cubs emerge from the bush. Suddenly you are aware of neither your pain nor your fatigue but only of the threat. Your attention has been hijacked. This, largely subconscious, parliament of modules, vying for control of our attention, decision making, memory and planning subroutines (and honed to survival perfection through natural selection) is a fascinating and very convincing view of the way our brain works. Add to this the illusion of vision as a window and its interlinking with the model building function of the brain and one is just in awe of evolution. No more the vision of some central controller, or spirit, sitting in your head watching the outside world and making decisions. Now we have a vibrant, multilayered, living brain in tune with and learning from its environment.

How do scientists know this? Again just 100 years ago they didn't know these details. The information comes from two main sources. Firstly, the tools we have now allow us to investigate things like how nerves connect to each other and how they use chemicals to transmit signals across the brain. Also, even more recently MRI scans have been made of the brain whilst subjects are performing specific tasks. With this, they can see increased blood flow to different areas of the brain, showing which areas are active during which tasks. The second main source is from work with patients that have specific, localised damage to areas of the brain, either through a very localised stroke or through some sort of head injury. With the current ability to directly identify the damaged area, it is possible to relate the function of that area to a behavioural or perceptual anomaly in the patient. See for instance the fascinating book by Oliver Sacks called *The Man Who Mistook His Wife for a Hat Stand*, one of many such studies in this area.

One interesting consequence of this modularisation of the brain is that you sometimes get overlap or leakage from one brain area to another situated next to it. If this occurs in the senses, it is called synaesthesia. There is the case of one notable composer who, as a child, thought that the reason the lights were turned down in a concert was to allow everybody to experience the 'colours' of the music. Also, the brain is very adaptable. If one area gets damaged, say in a stroke, an adjacent area may well start to take over the functioning of the damaged one, especially with training like physiotherapy.

Psychology

So now that we have an idea of how the brain performs its

processing, let us step back to the beginning of this chapter and re-look at the example list given of subsidiary aspects to our psychology (not complete by any means).

- A capacity for learning
- Curiosity
- A capacity for awe
- An appreciation of beauty
- A capacity for logical thinking
- A capacity for love

Where did these capacities come from and how did they evolve? The ability to learn things is an obvious survival asset but what would cause us actively to seek to learn? Evolution solves this by rewarding us for being curious, making us feel good when we discover things. It is like eating. We don't eat because intellectually we know we will starve if we don't. We eat because our body tells us we are hungry and because we like the taste and because we feel good when we are full. Evolution has given us this feel-good reward system. The same goes for sex. We don't make love because we deliberately want to leave descendants, we do it because it feels good. That is how evolution engineers survival. These things are strongly coloured by our emotional system, it is this that provides the reward and ensures that the things which are important for survival and reproduction are given strong significance. A capacity for awe is another example. Awe of the natural world and of its power to destroy us is a strong survival mechanism.

What about an appreciation of beauty, how could this be a survival or reproductive asset? Well what type of thing do we think is beautiful? I would bet that most people would include things like youth, vigour, symmetry, colour and sound (especially the human voice). Now think how valuable to survival these things are. It is obviously evolutionary better for instance to mate with a young, strong and well-formed person (symmetry is an indicator of not being deformed). Colours are often a good indicator that food is ripe for instance, and in some species (especially birds) colours are used to display during courtship. Also, for us, it is obviously important to be well tuned into the human voice, especially to its emotional content. Again, logical thinking probably arose for a number of related reasons to do with things like planning a hunt and anticipating the actions of others, obviously very useful tools for survival. And love probably has its roots in bonding and child rearing, especially so for humans where the childhood period is so prolonged.

So, as I said near the start of this chapter, these aspects of human nature did not evolve because they were good things in themselves to possess, they were designed as shortcuts to serve evolution's twin aims of survival and reproduction. But, having arisen, we (uniquely with our big brains) can now choose to use them as we see fit. Dawkins gives the example of contraception, a most unevolutionary idea, where we now have control over when we raise a family, severing the sex drive from procreation and leaving only the reward of pleasure. As technology improves, allowing more time for leisure, we can enjoy our other traits by indulging in music, storytelling, poetry, going out for a meal with friends or even just walking in the countryside.

Altruism

So much for selfish traits but where did altruism (i.e. goodness to others with no benefit to yourself) come from?

Darwin, writing in 1859, didn't know about genes, although he reasoned that there must be some method of providing heredity: the passing on of characteristics from parents to their offspring. But by the mid twentieth century we did know about genes and in 1976 Richard Dawkins pulled together many of the most recent ideas in evolution and genetics in his ground-breaking book *The Selfish Gene*. The title is of course a metaphor, he was in no way imparting consciousness or intent to genes, but a very powerful metaphor it was. It gave us a gene's eye-view of evolution and it worked. It was logical and it had the backing of mathematics in terms of the calculated frequencies of gene selection.

Prior to this, it was assumed that altruistic behaviour was selected for because it was good for the species but these new discoveries showed that natural selection didn't work at the group or species level. If a gene evolved that affected the behaviour of the individual in a way that increased the number of descendants produced, then that gene would spread well in the species (via those descendants). Natural selection works on the phenotype of genes, not on groups. This then gave us a problem, if altruistic behaviour could not be selected for at the species level, where did it come from?

Where one has to be careful, however, is the danger of transferring the metaphor of a selfish gene up to a selfish individual. Superficially it seems as if this type of selection pressure would make an animal selfish but there are a number of reasons why this is not the case. Firstly, one gene does not make an organ, like the liver or the heart. It takes a large number of genes working together in co-operation. Genes found early on that working together was a better way to get themselves into the next generation than working individually.

This works at a number of different levels. If we look right down at the level of an individual cell we find a division of labour between the different parts of the cell, with some genes coding for the proteins of the cell wall, others coding for the DNA repair proteins, others coding for the protein building complexes themselves and so on for all the myriad of special functions within the cell. Specialisation, division of labour and co-operation is the order of the day. As single cell organisms became more complex through evolutionary time, they found great payoffs from genes that worked together. A particular case in point is a cell's energy storing bodies, which are called mitochondria. It is widely believed that these tiny bodies were once separate cells in their own right, then at one point in the distant evolutionary past (sometime around 2,500 million years ago), they merged with another cell and the two cells found it extremely beneficial (for both their sets of genes) to work together. This merged cell had a great survival advantage and it became the ancestor of all life that is made up from complex cells. Every cell in your body is powered by these tiny mitochondria, as is true of all animals and plants.

At the next level up multicelled organisms appeared (around 1,400 million years ago). Individual cells found that it could be very useful to work together and it paid them to specialise at individual jobs, for example digestion, respiration, sensory, movement etc. You even get cells that deliberately commit suicide for the good of the organism. This works because the genes for that cell are not lost but are carried on in the other cells in that organism.

But how could altruistic behaviour benefit the individual? In the 1960s the biologist Bill Hamilton showed mathematically how aiding a relative, especially a close one, can make evolutionary sense because they carry a large number of the same genes. Again, here we see a gene's eye view of natural selection. In fact, this theory became known as 'kin selection'. So if a gene arose that affected the behaviour of an organism such that it would assist its relatives when they needed it, then that gene will quite likely get selected because the relatives that survived due to the sacrifice of the individual would also be likely to carry that particular gene and hence would pass it on. This can be seen most clearly in things like ant colonies, where an individual ant is more closely related to her sisters than to any offspring she would produce. So you get specialisation of job function in an ant colony into worker, fighter etc, neither of who lay the eggs, that is the queen's job.

In other animals too, such a gene that assisted relatives (identified by smell for instance or just statistically likely because of living together) would prosper in the population. In extreme cases it would even be cost effective (in evolutionary terms from that gene's point of view) to lay down your life to save the life of two siblings, each of whom has a 50% chance of having that gene. For cousins it would have to be for saving four because they only have a 25% chance of having that gene. The full mathematical theory was worked out by Hamilton. And in the book *The Selfish Gene* Dawkins gives a much more detailed account than I have space for here.

So, we get co-operation at all levels of biology and we get kin selection or behaviour to care for and assist close relatives arising naturally through evolution. But it does seem that there is still a question to be answered about altruism or helping strangers at a cost to yourself. Darwin too had problems with this and, while he had the suspicion that the answer was somehow linked to us being social animals, living closely together in groups, he never fully understood how such behaviour could come about and he left it for future generations to complete his theory. This brings us to two new discoveries: 'game theory' and 'mirror neurons'.

Game theory is a branch of mathematics that looks at what strategies are most likely to give the best payoff for a given game with a given set of rules. The analysis of one particular game, known as 'the prisoner's dilemma', has given us an interesting perspective on how people interact and work together. The classic form of this dilemma is where two criminals have been arrested and are being interviewed separately. If they both keep quiet then they will both get a short sentence. However, if one informs on the other but the other keeps quiet then the informer will go free and the other will get a long sentence. However, if they both inform on each other they will both get a medium-length sentence. Obviously the best result for either one would be to go free but that involves informing on your mate. And what if he, following the same logic, informs on you? Then you would both have to suffer the medium-length sentence. But would you trust him to keep quiet? The best mutual result is if they both keep quiet (honour among thieves) and both do a short sentence.

A different version of this game has been tried a number of times with real people taking part. In this version the prize is money (as opposed to going free). Two people play and an umpire pays out any money won and collects any fines if required. Each person has two cards, one which says 'co-operate' and the other says 'defect' (co-operating is helping the other person, equivalent to the prisoner keeping quiet to help his mate). They both play one card, face down, at the same time. If they have both co-operated then they both get say £100. If they both defect then they both get fined £10. But if one co-operates and the other defects, then the defector gets £200 and the one who co-operates gets fined £50. (The actual amounts don't really matter as long as their values relative to each other remain the about same.)

Again, you can see that the best individual solution is if you defect and the other person co-operates, you get £200. But can you trust him to co-operate? If he defects too you both get fined £10. You both know that if you both co-operate you both win £100; not as good as the £200 but better than a fine. But if you co-operate and he defects, you get fined £50. Again, can you trust him to co-operate?

When running this game as one-off encounters between strangers they found all outcomes were tried. But when they let two people play it many times, they found that people would begin to trust each other and they started co-operating. If there was the occasional defect then the other person would pay them back by defecting the next time and then they would often get back to co-operation.

Later, computers became involved. There was a competition to write the best program to play the game and the programs were all pitted against each other to find a winner, i.e. which program had the best strategy. There were some that just co-operated all the time, some that defected all the time, and a range of strategies in between. A good strategy was one called 'tit for tat', which copied the previous move of its opponent. However, when playing against another 'tit for tat' program they would get locked into a never-ending cycle of alternating co-operation and defection. The best strategy seemed to be a version of 'tit for tat', which allowed an occasional defection without retaliating.

This has become known as reciprocal altruism. You would be good to others as long as they are good to you and you would probably give strangers the benefit of the doubt on first meeting. For animals living in small groups it is fairly easy to keep track of other individuals in the group, remembering who was good to you last time and who wasn't. Grooming each other is a good example; one animal spends its time removing parasites from another and would then expect the same service in return. As groups became larger then it takes a bigger, more sophisticated brain to keep track of all the other individuals and who owes what to whom.

Another problem is how do we know we can trust another individual to co-operate and not defect? And, in reverse, how do we convince others that we can be trusted to co-operate? How do we convince them that we are not lying? Well, think how you react if you lie to someone. Your pulse rate increases, you perspire slightly, your pupils contract and you may well go a bit red in the face. You can't help it, it's an automatic response. We wear our hearts on our sleeves to advertise our trustworthiness in a way that can't be faked. Any genes that arose which made us behave in that automatic way have obviously been selected for. Animals that co-operate obviously do better than ones that don't, as borne out by game theory.

We need to be gullible as a child to soak up all the learning needed to survive in our complicated world but by the time we reach adulthood we also need to have learned how to assess others. We are all excellent amateur psychologists, good at reading people and detecting liars and cheats. Given the trusting nature of our society the odd freeloaders can flourish but not too many of them or co-operation breaks down. An excellent book examining these issues is *The Origins of Virtue* by Matt Ridley, it is highly recommended.

The other new discovery recently is 'mirror neurons'. Using the latest tools to track brain activity, scientists can see which groups of cells are active during which particular tasks. They discovered a particular group of cells that fired not only when the individual being studied performed an action but also when that individual watched someone else perform the same action. It was as if the watcher was mentally performing that action too. The watcher is able to see things from someone else's viewpoint, she then has an internal idea of what the other person might be thinking and feeling, an empathy with that other person. This is extremely useful in highly social animals, not only do you get an idea of what the other person is really thinking but it helps you to be able to copy their actions – learning through imitation is one of the basic elements of culture.

Empathy has been well documented in chimps, it has even been reported in mice. Social animals virtually require it when the groups get bigger. It is also useful for reciprocal altruism; you need to have an internal model of other minds and their viewpoints in order to decide on whether to trust them or not. Most of this modelling and empathy processing is of course largely subconscious, all you are consciously aware of is a feeling of trust or not for someone, an empathy for them, often difficult though to put the reasons for it into words. Words usually come afterwards as a kind of rationalisation for the feelings you have.

The Illusion of Self

So the idea of an 'I' sitting inside watching the world and making rational decisions is obviously an illusion, albeit a very powerful one. Again, this 'I' is one of the models we create of the world, just as we create internal models of our friends. The great Scottish philosopher David Hume realised this even without the scientific tools we have today. Writing in the mid 1700s he pointed out that we think we are the same individual we were say five years ago even though we've changed in many ways since then. He also said, 'Man is a bundle or collection of different perceptions which succeed one another with an inconceivable rapidity and are in perpetual flux and movement.' He used the analogy of a commonwealth, which is not a single thing but a loose, changing association of related elements. (The idea of a constant entity made up of different and changing parts reminds me of the farmer's wife who said, 'I've had this old broom for years. It's had a few new heads and once it had a new handle but this old broom has done me well for years.')

The French mathematician and philosopher Rene Descartes, writing in the early 1600s famously said, 'I think therefore I am.' This was really the beginning of modern philosophy, when he refused to accept existing theological dogma and tried to derive what is true from first principals. Unfortunately in doing so, he fell for the illusion of the self as an internal 'I'. He compounded this when he went on to identify the separation of mind and matter, an understandable conclusion given the ideas and knowledge of his time but one that set philosophy on a wrong track for many years. Even though thinkers like Hume saw things more clearly, Descartes' influence can still be discerned today.

We know from doctors and psychiatrists that someone's personality can completely change with the onset of an illness such as dementia, a stroke, a tumour or even just a head injury. It has been well said that we are all just a fall or a burst blood vessel away from being someone else. Who we are at any one time is a function of the current parliament of modules (brain subsystems), our memories, our brain chemistry and the models we build and use to interact with the world. All crafted by evolution and experience; nature and nurture.

This incredibly useful concept of 'I' (our interface with the outside world) is to a large extent a function of language. Other animals, especially primates, may well have a similar internal model of themselves. This has been well demonstrated by the ability of some of them to recognise themselves in a mirror. But it is language that has cemented this illusion into our thinking. Not only because we talk and think about ourselves as 'I' but also because (as we saw in the section above on empathy) words usually come after feelings and after all that subconscious processing, as a kind of rationalisation of our internal state. Almost as if what we think and feel is a story that we tell ourselves after the fact.

So there is no internal 'I' inside us and there is no separation between mind and brain. We are a bundle of perceptions, a parliament of modules, a largely subconscious model building empathic ape with an aptitude for language and logical thinking.

Free Will

Another problem for philosophers through the ages has been where does free will come from? This seems to become even more of a problem if we see the world from a scientific, deterministic viewpoint.

Daniel Dennett, in his excellent book *Freedom Evolves*, examines this problem of where freedom comes from if the world is fully deterministic. That is, if we imagine being able to know where every particle in the universe is and how it is moving then we would be able to work out what is going to happen in the future, in fact from this point of view the future would be completely determined, hence we would not really be 'free' to act because everything we did would be pre-determined.

He points out two problems with this view. Firstly, there is a question as to the possibility of ever doing this because you would have to deal with things like chaos and the uncertainty principle in quantum mechanics (i.e. that it is impossible to determine both the exact position of a particle and its velocity at the same time). The second difficulty is that any model of the universe would itself be part of the universe and so would have to model itself, and to model itself modelling itself... ad infinitum.

So, although we can consider the universe to be deterministic, with effect following cause for instance, it is impossible to know what will happen in the future. Now obviously organisms in the world could not know what is going to happen but they will certainly survive better if they avoid predators and other dangerous situations. So, they do best when they take in information about the world and make decisions based upon that information – run, fight, eat or mate. Simple instincts can do very well for many creatures. Stephen Jay Gould tells the story of when he was a youngster in the US navy, an older seaman gave him the following advice: 'If it moves, salute it. If it doesn't, paint it,' a good way to keep your head down and survive without getting into trouble. Gould went on to show how simple, hard-wired instincts serve many animals very well, especially where the cost of running a large brain is prohibitive.

However, the more brain power you can bring to such decisions the more freedom you have to choose a good path. In other words, as per the title of Dennett's book, freedom evolves, even in a deterministic world.

Slightly more pragmatically, American philosopher Patricia Churchland sees free will as a necessary social requirement. We need to be able to hold people responsible for their actions. Sometimes, for instance, people genuinely can't help themselves if they commit a crime but most of the time we do hold people responsible. She suggests moving the debate over free will away from metaphysics and towards the neurology of self-control and the ways that this could be compromised. Unlike free will, self-control is a concept that can usefully be applied to other animals and hence brains in general. It is something that increases as the organisms mature. So, does this view cheapen us? She emphatically says no: –

> *The beauty, intricacy and sophistication of the neurobiological machine that makes me 'me' is vastly more fascinating and infinitely more awesome than the philosophical conception of the brain-free soul that somehow, despite the laws of physics, exercises its free will in a causal vacuum. Each of us is a work of art sculpted first by evolution and second by experience.*

> **Patricia Churchland**

Conclusion

> *My soul is a hidden orchestra; I know not what instruments, what fiddle strings and harps, drums and tambours I sound and clash inside myself. All I hear is the symphony.*

> **Fernando Pessoa**

So, our brains model the world and do so well below the level of conscious thought. We think we 'see' the world as it really is but what we are really experiencing are the models, the symphony, the output from that parliament of modules in our brain. And these mental models, our ways of thinking and our behaviour have all been designed and modified by natural selection to give our genes their best chance of surviving and reproducing as they strive for the future.

I'm going to end this chapter here. Both neurology and evolutionary psychology are huge subjects in their own right and there is room in this book only to scratch the surface of all the new developments in these fields. I hope, however, I've stimulated your interest and I believe I've covered enough to present a flavour of these fascinating topics and to meet the twin themes of this book: to provide an understanding of where we came from (and in this case where our behaviour comes from) and to demonstrate some of the reality of the world beyond how we perceive it through our senses. All this is new findings over the past 100 years or so, indeed a lot comes from just the past few decades. Science shows us a world that is stranger, more wonderful and awe inspiring than any fiction (or ancient scripture written hundreds of years ago) and if we ignore what it tells us we are, at best, deluding ourselves. Our brains make models of the world; we owe it to ourselves to make those models as well informed as possible.

Chapter 11

The Scientific Method

Trying to understand the way nature works involves a most terrible test of human reasoning ability. It involves subtle trickery, beautiful tightropes of logic on which one has to walk in order not to make a mistake in predicting what will happen.

Richard P Feynman

There have been incredible advances in our knowledge of the world over the past 200 years or so and indeed that very progress seems itself to be accelerating in recent times. But, on hearing about new, often counter intuitive ideas, the question often asked is 'How do they know that?' As identified in previous chapters it has been the use of tools that has enabled many of these amazing, almost incredulous discoveries, providing knowledge about realms that our senses alone would never even be aware of, magnifying the minute, reaching out to the furthest stars and 'seeing' via different spectrums.

Two other significant factors, however, have greatly contributed to these advances. The first is the Lamarckian nature of culture (again previously described) which, by its successive refinement of ideas and of the methods of building new and improved tools, inherently provides exponential development.

The second significant factor, however, and arguably the more important, has been the development of the scientific method, which is now the predominantly accepted way in which knowledge is obtained and, more crucially, verified. The method consists of the following six steps: –

1. Produce a theory to explain the existing facts.
2. Use that theory to make predictions.
3. Perform experiments to test those predictions.
4. Ensure those experiments are repeatable (by yourself and others).
5. Refine (or abandon) the theory in the light of the experiments.
6. Keep repeating the whole process.

Another important aspect to this whole process is peer review. Write-ups of new discoveries or new theories are only accepted for publication in the main scientific press after they have been sent out to other leading scientists in the same field and their reviews of the work prove to be favourable. That is not to say these other scientists actually reproduce the findings at this time, only that they agree that the science behind the findings is considered sound and the inferences and conclusions are logical and mathematically rigorous.

A theory is only a good theory if its predictions can be verified and it will only become widely accepted over time if other people can reproduce the same experiments, make more predictions and test those too. Science, at its best, is a collaborative, Lamarckian, enterprise.

Someone once said to me something along the lines of, 'I've thought about what you said about a scientific theory, that it must explain all the facts and fit the available evidence and I've come up with one that supports my faith.' Well of course, he had missed out the most important part of the scientific method. Anyone can come up with a theory that fits the facts, for instance I could claim that I have a special powder that I spread around at night that keeps tigers away and then I say, 'Look, it fits the facts, it must work because there are no tigers here.'

A scientific theory must be one that is capable of being falsified, if not then it is not a scientific one it is just guesswork. To be scientific you must be able to use the theory to make predictions and to be able to test those predictions, not just once but repeatedly. If a theory survives repeated tests then it is a good one but scientists will still keep an open mind and do further tests as new tools and ideas become available.

If you can't use the theory to make predictions that can be tested then it is not a scientific theory, it is just speculation. The part my friend missed was using his theory to make predictions and then repeatedly testing those predictions. This is the major difference between science and religion, scientists are constantly trying to falsify their own theories, testing them in as many different ways as they can think of to see if they still hold true. A theory may work well in 100 different tests but if on the 101st test a small fact pops up that the theory does not explain, then either the theory needs amending or it is wrong. Religion on the other hand relies on given texts that are not allowed to be questioned (even though they have probably been handed down over the years and translated many times!).

If someone claims something to be true, the first thing you should ask is 'Where is your evidence?' Then you should ask 'What has been done to test it?' and crucially 'Have other people been able to reproduce these tests under controlled conditions?' Take the tiger powder claim, for instance; some very simple tests spring to mind. I could stop using the powder and see if tigers appear, or better still, I could take the powder to where there are tigers in the first place and see if it works there. In fact, it is beholden on me to do everything I can to prove my own theory wrong! Evidence is all.

When generating theoretical ideas scientists should be fearlessly radical, but when it comes to interpreting evidence we should all be deeply conservative.

Peter Coles (Astrophysicist)

One problem to look out for here is that when the non-scientific press get hold of a story, they often blow it up out of all proportion. In the interest of a good story new discoveries are often portrayed as overturning the established views, proving everyone else wrong etc…when all that is really happening is that an existing theory might need to be modified slightly, often in a way that experts had suspected it would need amending in the first place. (That of course would not sell as many papers.)

Another problem with the popular press and especially with TV is that they always think that they have to show two sides to every story. Either because they think that they need to be seen as being fair and even-handed or, more mischievously, to create controversy. They do this even when there is no real other side to a story, creating doubt and confusion where none ought to exist. There are plenty of cranks around so it is easy for the media to do this.

We all like to have explanations for things, it's the way our minds work, even if we don't follow all the steps in a line of reasoning ourselves it is comforting to know that there is an explanation for something and that someone does understand it. With the success of science and the scientific method in recent times, there has been a general public acceptance that anything that looks as if it has some sort of scientific backing must have a large degree of credence attached to it. You see this all the time in the advertising industry. Men in white coats, scientific looking equipment in the background that you are supposed to assume has been used to test the product, scientific sounding words like hypoallergenic being thrown about: in short, science sells.

At the same time, there is a harking back to simpler times, natural products and ancient wisdom. Anything that combines these two themes has strong appeal and therein lies the danger. There are all sorts of odd theories around, many that have been in existence for a long time but that are now adding the illusion of scientific explanation to their presentation, often with copious literature to back them up. For instance you have astrology, homeopathy, crystal therapy, spiritualism, intelligent design, alternative this and alternative that, the list is endless and the more they include words or word endings like 'therapy' and 'ology' the more believable they sound.

Take homeopathy for instance – practitioners set themselves up like doctors, their potions are delivered in medicine-like bottles and their literature is huge. The basic premise of their practice, however, is that they find something (a chemical, herb or mixture of some kind) that they think will help cure somebody and then they dilute it in water often by a factor of 100. Then they take that dilute mixture and dilute it again by a factor of 100 and they keep doing this up to about 30 times. Presumably on the premise that the more dilute the mixture the stronger the effect. (Be careful next time you make a cup of coffee to keep you awake – if you only use just one crystal of instant coffee in your cup you will be flying!!) They call this the scientific sounding name '30C'. This actually means they have diluted the mixture by 100 to the power of 30. If you work it out that means it is less dilute than having one molecule of the original substance in a volume the size of our solar system! The likelihood of anything active being left in their little bottles is next to nothing. But then when challenged on this they come up with theories like, 'Water has a memory of what was previously dissolved in it'!

The main problem with most of these theories is that they will not allow their methods to be put to rigorous scientific tests. The human body is remarkable in its ability to heal itself, its immune system is extremely efficient at fighting infections, indeed humans wouldn't have survived until the present day without this being so. Add to this the power of the placebo effect and undoubtedly some people will recover after undergoing an alternative therapy. The only way to separate out all these factors is by testing with large-scale, well-designed double-blind experiments.

Take astrology; there is a story about a young journalist who had just joined the staff of a small provincial newspaper. Being the most junior employee, he was given as one of his jobs the task of writing the weekly horoscope. He had at his disposal all the past horoscopes and was told to create new ones in a similar style. This he did for a number of weeks until at last he got really fed up with not doing real journalism. And so one week, under one star sign he wrote something like 'all the troubles and sorrows of yesteryear are as nothing compared to what is going to befall you today'. The switchboards were soon crammed with calls coming in, worried people asking what they should do to be safe, should they stay at home or go out or what?

We are easy to fool, we trust people. If someone tells you

something, you will likely believe them. This is part of our evolutionary nature; we are social animals living in groups and information passed between people is an important way of learning about the world. Bad science hijacks this trust and there is a lot of it about, building up its own memeplexes in a similar way to religion. We desperately need to be aware of this and we need to be able to discriminate between good and bad science. There is no easy way to do this, especially in this age where we are constantly bombarded with information from all areas about all sorts of things. The only way is to keep reading about a particular subject from a number of different sources. If you rely on only one source, you will undoubtedly get a biased view. As well as that, some sources can be viewed as more trustworthy than others, for instance you may well regard the BBC as more accurate than say a tabloid newspaper. More trusted still in scientific matters are the main science journals, which publish only articles that have been peer reviewed. We are all gullible, there is an old joke where one person says to their friend, 'I never believe anything unless I hear anecdotal evidence that it is true.'!

In this book I have mentioned or quoted from a number of people, usually from a scientific background, whom I have come to trust and I would thoroughly recommend you to read any of them. Again, however, read round the subject, read what other people have written in criticism of their point of view, then read other criticism of that criticism and then make up your own mind, but do so based on the evidence, not on hearsay or newspaper articles.

Another thing that can influence people's opinions is coincidences that have happened to them. These often have the appearance of fate, destiny or of some hidden significance about them, but genuine coincidences do happen. In fact, there are so many things that could happen that would be called a coincidence if they did, that, statistically speaking some are likely to happen. For instance, I once bumped into my sister in London at a time neither of us lived in London (I lived in Southampton and she in Bedford), also we didn't know the other was going to be there. Wow, what are the odds of that?! But when you start to think about it, a) the meeting occurred in a train station that we would have to use to get to our parents' Bedford home and b) what if it had been one of my brothers instead of my sister or what if it had been a friend from Bedford or indeed from somewhere else, each just as unlikely on its own but there are so many meetings that could have occurred that the odds against something happening are nowhere near as outlandish as you first might think. There isn't always a cause, much as our brains would like there to be.

Science and the scientific method have explained an incredible amount about the world and the universe over the past 200 years but that doesn't mean we now know everything; there is still much to discover and quite possibly some things that can never be known. The important thing is to stick to the scientific method and not revert to wild speculation or the supernatural just because we don't yet know the answer to some things. We are working on it.

Some of the important areas of the picture that we don't yet know about in detail have been mentioned before in other sections of this book. They include reconciling relativity with quantum mechanics and identifying the true nature of particles – are they solid particles, waves or even tiny loops of string vibrating in 11 dimensions? Also, the latest results from satellite data say the universe is made up of only 4% ordinary matter as we know it, 21% dark matter and 75% dark energy. What on earth are dark matter and dark energy? There is nothing mystical implied by these terms, physicists are just using the word 'dark' because they don't yet know what it is.

Dark matter is inferred to be there because we observe that stars are revolving round their galaxies too fast for the gravity of the visible mass of each galaxy to keep them from flying off. Therefore, there must be some extra mass and gravity somewhere keeping the stars in orbit around their galactic centres. One candidate for this extra mass is weakly interactive massive particles (WIMPS). These are particles, predicted by theory, but which don't easily interact with normal matter hence they are difficult to detect (we are working on it). So 'dark' just means we can't see them yet.

Also, dark energy means energy we predict by theories and

measurements but we can't yet identify the true nature or source of it. In this case, it is predicted because we measure that the expansion of the universe seems to be increasing, whereas before we had thought that gravity would be slowing it down. The best theories so far seem to be that dark energy is some sort of positive energy of empty space. However, these are very difficult measurements to make, based on long chains of reasoning and distance measurements of remote galaxies, so the possibility exists that the theory of dark energy might not be needed if our measurements improve, although currently it is our best guess that dark energy is real. (The reason they talk about dark energy as making up part of the universe is because, as Einstein showed, energy can be considered as being equivalent to mass.) We live in interesting times, with an explosion of current knowledge but with much still to uncover.

We are all uniquely privileged to be small parts of the universe that are actually aware of the universe itself, and as such we owe it to ourselves to find out as much about the universe as we can. Our main tool in this respect is the scientific method. To be seduced by woolly theories, mysticism or religious hocus-pocus would be self-deception of the worst kind and a betrayal of our stellar origins. We have only just started out on our voyage of discovery, the amazing picture of nature roughly outlined in this book is being revealed by scientists bit by bit, like an art restorer carefully cleaning an old master. What stunning hues, what subtle textures and what awesome details are yet to be uncovered from nature's self-portrait and revealed for all to see in the galleries of knowledge?

Chapter 12

The Future

The empires of the future are the empires of the mind.
Sir Winston Churchill

A question often asked is 'Are we still evolving?'

This is a fascination question, not only because it sheds light on us as humans, but also because it demands a definition of our terms, it requires us to understand more fully the nature of evolution. In chapter 6 we looked at how evolution works through natural selection and at some of the current misconceptions about evolution. Here we will look more closely at some of its implications.

Many years ago, as a child, I watched an early science fiction episode of a TV series, *The Outer Limits* I think it was called. In this episode a scientist had built a machine that he claimed accelerated evolution. He climbed into it and turned it on. And, as his assistant watched, the scientist's head grew bigger with his brain expanding, also he grew a sixth finger on each hand and by the time he got out of the machine he had extra powers like telepathy and the ability to move objects with his mind. Of course, then he overreached himself and tried to dominate everybody (obviously his ethics hadn't improved much!) and in the end he got his comeuppance in some way I can't remember, presumably for trying to play god or some such moral. A good yarn, with a touch of darkness thrown in, typical of the series and of the time. Again too, see the confusion of 'evolution' with 'progress' *[discussed at length in chapter 6].*

The program assumed that our descendants will have bigger brains as if there were some in-built direction to evolution. As we have seen, evolution isn't 'going' somewhere, it is just change in response to local situations, it has no aim, no foresight.

The next obvious thing to point out is that individuals don't evolve. It is their descendants that may or may not be different. So, the question 'Are we still evolving?' must refer to the human race (species Homo sapiens) as a whole. But here we meet another complication – a species can't suddenly change, a species is made up of individuals. Also, as we saw in chapter 6, natural selection works at the genetic level, selecting for gene variations that aid survival and reproduction. Perhaps a better way of looking at a species is the idea of a gene pool that belongs to the species as a whole. This would be all the genes that go to make up an individual of that species together with all the variations for each of those genes. Many genes are so fundamental in their way of working that there would be few if any variations of them that would still do their job, but others (like genes for eye colour or body hair length) may well have a number of workable variations. The genes in this gene pool reside in the current population of individuals within the species. These are temporary resting places as the genes try out different combinations of working together to find the best ways for flowing into the future.

So is the gene pool evolving? Well the short answer is yes. New variations of genes are occasionally created by chance mutation, and, if they don't prove fatal to the individual, then they will take their place in the gene pool. Bad variants get weeded out pretty quickly because the bodies they make are not up to the rigours of surviving and reproducing. Good ones, however, will spread if they do better than any other variants of that same gene. Many variants though will be evolutionary neutral.

The next question is how long will it take before a good new variant of a gene resides in all individuals of that species? In biological terms, this is known as the gene becoming fixed. You can quickly see that this depends on how many individuals there are in that species. If you have an isolated population of few individuals and if that gene variation gave good survival or reproductive success then any descendants possessing that variant would do much better than any without it and they too would pass it on. Soon all surviving members would possess it. If, however, the population is large (like the current situation for Homo sapiens) then it could take millions of years, if it even happened at all.

If a population of a particular species is very small we say that species is passing through a bottleneck. This could be caused by a catastrophe of some kind or perhaps through a group of individuals becoming geographically isolated from the main population or, quite likely, an illness or virus decimating the population. In this case, only those individuals whose immune system could cope with that illness would survive, presumably because they possessed a particular variant of one of the immune system genes which others of the population didn't have.

As mentioned in chapter 8, our immune system is one of the most complicated parts of our body. It has to be able to recognise and fight millions of different viruses and bacteria, most of which it has never seen before. It also remembers those things it has met before, enabling it to fight them off much more easily next time, and it has to distinguish between our own cells and invader ones so as not to turn against its own body. The ability of an individual to defend itself in this way is fundamental to the development of a larger, multicelled body and must have been one of the earliest evolutionary developments. Our own immune system has a number of different parts to it and is controlled by many of our genes. As we all know, however, some people are more susceptible than others to various illnesses and there must be many variations of the immune system genes in our gene pool.

Individuals dying due to illness is classic natural selection, leaving only the fittest individuals to reproduce, hence only successful genes being passed on. This is especially the case in a small population. When such a bottleneck occurs, many gene variants are lost and of course, there is the danger of the population becoming too small and of inbreeding problems occurring. But it will often be here that a whole selection of gene variants become 'fixed' due to their alternant variants being lost. A lot of that 'fixing' will be evolutionary neutral just by the coincidence of being in the right place at the right time.

If there is one place that our own gene pool is currently evolving it is probably within the immune system genes, with local populations suffering losses while some of their number are able to fight off the illnesses (albeit this is now complicated by the use of medicines). One recent example of this was the 'Black Death' of the medieval period, which originated in the Far East and then spread through Europe. In 1348 it killed about one third of the population of England with huge economic impact. Immune gene variants which were able to resist this would obviously survive and then be passed on.

However, there is little likelihood of any currently arising variants becoming fixed in the whole population of the world simply due to our numbers. Note also (as pointed out in chapter 6) there is no sense of direction or progress here. 'Fit' here only means fit at fighting that disease or virus, it is not used in any sense of 'goodness'. Also, these things often come in cycles with a disease disappearing when most of the population become immune to it but reappearing generations later when that immunity is lost because it is not being used.

Interestingly, although our current population is so large, and on the surface so diverse, our own gene pool is incredibly homogenous or uniform. It has been said that there is more genetic diversity between two gorillas from one troop in Africa than there is between say a given European and a Chinese person. Genetic evidence on these lines, backed with archaeology, point to the fact that Homo sapiens as a species went through a very small bottleneck in their history.

Our species was anatomically human 200,000 years ago but from a study of artefacts, we became behaviourally human between 50,000 and 80,000 years ago. From the analysis of our genes, it is estimated that our numbers reduced to around only 15,000 individuals (we were very nearly wiped out altogether). This occurred probably about 70,000 years ago and seems to be concurrent with a period of huge volcanism that put so much ash into the sky round the globe that it reduced the sunlight to such an extent that it started an ice age which lasted perhaps 1,000 years.

During these tough times evolutionary pressures would have been strong and there would have been selection for things like a capacity for planning and for logical thinking. The ability to make and use tools in hunting would have spread quickly due to our newly developed language skills. Indeed, it seems likely that there would have been a strong positive feedback between language, culture and intelligence at this time increasing the selection pressure.

An interesting point here is the difference between hard times and bottlenecks. During general widespread hard times, it is the pre-existing variation within the gene pool which gets selected for, as the less fit genes succumb to the rigours of the environment. This does not necessarily mean that these selected genes will become fixed because the population could still be quite large. However, with bottlenecks, any gene variants surviving will probably become fixed, be they fit or not. It will often be the case of course that hard times and bottlenecks occur together.

Also, the difference between pre-existing variation and newly arising variation of genes should be highlighted. In an established species there will inevitably be a lot of pre-existing variation within the gene pool and it is this that mainly gets selected, either for or against, during hard times or bottlenecks. When a new variant arises in an individual and where this is beneficial (a very rare occurrence) it takes a long time for it to spread because, of course, genes don't spread sideways, they can only be passed down to descendants. A fascinating example of a recent mutation to one of our own genes is a variant that allows adults to tolerate drinking milk. Normally in mammals milk drinking tolerance is lost during late childhood. Most adult humans in the world can't tolerate milk but in Northern Europe more than 90% of adults have the variant that allows them to drink it. This variant arose in Europe in Neolithic times (around 7,000 years ago) and has spread widely here, presumably because it conveys great survival advantages, a safe, all-year-round source of food, calcium and vitamin D. It is nowhere near being 'fixed' in our species, however, due simply to the size of our population around the globe.

Let's return, however, to the original question 'Are we still evolving?' Often when this question is asked there is a hidden question behind it... 'Are we currently becoming more intelligent?' The answer here is emphatically no! Superficially it looks as if we are but that is an illusion created by culture. We are learning more about the world all the time and undoubtedly we know more than our cavemen ancestors (and perhaps we are more 'civilised') but that is accumulated knowledge and culture, not intelligence. We are genetically virtually identical to those individuals who were lucky enough to survive the bottleneck 70,000 years ago.

Timescale is important in this discussion. Obviously there has been about a six-fold increase in brain size since the common ancestor we share with chimps, about six million years ago. As discussed in chapter 6, evolution can sometimes be pushed in a particular direction, due to predator pressure or perhaps sexual selection. Also, the delayed adulthood of Homo sapiens allows more time for the brain to continue growing, which again as we saw before could well be controlled by just a few genes. But we also know (from chapter 6) that there is no such thing as in-built direction within evolution and that any idea of 'progress' is also an illusion (fed by wishful thinking and our overblown view of our own self-importance).

For movement in a given direction (such as brain size) there would have to be a) some sort of selection pressure and b) a bottleneck of small population size. Neither of these requirements presently exists. There is no current mechanism that prevents people of small brain size, or low intelligence, either surviving or reproducing and certainly there is no current bottleneck of population size! In the past both conditions were probably true, indeed it may well have been that our ancestors such as Homo erectus out-competed or even killed outright other pre human lineages and certainly large brain size was a survival asset. Also, it may well have been sexually selected for, with intelligence being an attractive quality. Dominance (due to intelligence) in a group may well have restricted breeding opportunity too, as was the case with the Aztecs of South America where only the king and anyone he gave favours to had mating rights. The case of the Aztecs itself would not have affected the whole gene pool of course but if a similar social set-up occurred in the past during a population bottleneck then that could have had an effect.

So, the (cherished) idea that we are slowly becoming more intelligent is not true in the current sense. In the past it obviously was but now there is no selection pressure in that direction and anyway the population is far too large for any such changes to spread well, let alone become fixed. You don't have to go far back for selection pressure to become important. Just 500 years ago a baby born in Britain only had a 50% chance of surviving to reproductive age, now such a baby has a 99% chance. So, you could argue that one way our gene pool is altering is that many gene variants that in the past would have been selected against are now being allowed to survive. You also don't have to go far back for the population of the world to be much smaller. In 1900 there were 1.6 billion people, in 1952 (when I was born) there were 2.5 billion, today we have over 6 billion, an incredible increase.

Predicting the future is really just guesswork but, in trying, one of the best guides we have is the past and a large shaping force for life on earth in the past has been catastrophes. As we saw in chapter 6 there have been at least five major catastrophes to life on earth, each causing mass extinctions. These are listed below, giving the name of the geological period and the number of years ago in millions (mya): –

- Cretaceous, 65 mya
- Triassic, 208 mya
- Permian, 245 mya
- Devonian, 360 mya
- Ordovician, 438 mya

The Cretaceous one, in which the dinosaurs were killed, was caused by a meteor. The other ones may too have had the same cause but there are a number of other possibilities, including massive volcanism or even a nearby supernova. Many people believe we are currently living through a sixth major extinction event because the background extinction rate of plants and animals now is 1,000 to 10,000 times higher than normal. This they think is being directly caused by us, via direct killing, habitat destruction, pollution and global warning. Another mass extinction event from natural causes in the future is inevitable, given enough time, but we do seem to be bringing one on ourselves. With our current population levels, our overuse of natural resources and our pollution of the planet it is easy to predict that if we don't change our ways, the future will hold many problems for us, like huge numbers of refugees, water shortages, wars and famines.

If we disappeared tomorrow, the planet would be in a good state but if the current extinction of plants and animals goes on, the future for humanity looks bleak. The environment is fragile and we need it to support us; see the following from the great biologist E O Wilson: –

> *Humanity did not descend as angelic beings into this world. Nor are we aliens who colonised earth. We evolved here, one among many species, across millions of years, and exist as one organic miracle linked to others. The natural environment we treat with such unnecessary ignorance and recklessness was our cradle and nursery, our school, and remains our one and only home. To its special conditions we are intimately adapted in every one of the bodily fibres and biochemical transactions that gives us life.*
>
> **E O Wilson**

If, however, we do solve these problems, somehow obtaining unlimited cheap energy and food for all, then other options for the future of humanity do open up. We can already inspect the genetic make-up of fertilised human eggs after they have gone through just a few divisions. We can test for some inherited genetic disorders and of course the sex of the baby to be. This capability is going to expand exponentially over the next few decades and the costs of such tests will come right down. It will, in the not too distant future, be not only possible to identify a huge range of genetic illnesses but also to select positively for a number of advantageous traits such as strength, longevity, musical ability, intelligence and others. These of course would just be potential abilities for the child, the right educational and emotional environment would be needed to build on these latent possibilities.

Here we enter the moral minefield of designer babies. How far do we want to go down this route? What parent, given the choice, would not want their children to be free of genetic illnesses? Yet we remember all too clearly the eugenics nightmare of Hitler's Arian race and his persecution of anyone not conforming to his vision. It is a discussion that we as a society will need to engage with in the coming years as science's ability outstrips our current social mores.

Will this change humanity? It would certainly be good to eradicate some diseases if possible. Again, though, we hit the numbers problem. Would everybody on the planet have access to this technology and would everybody want it? Would all the choices, if made, be in the same direction? We will certainly, in the future, have the ability to do some of these things and whatever changes we do make will affect the gene pool. But the pool is large, it would take a long time and everybody moving in the same direction to have any significant effect.

Artificial Intelligence

The other aspect of the future I want to consider is computing. This was touched on briefly in chapter 8 but here I want to go into more detail because I think this will be one of the major developments in the future. Again, if we use the past as a guide, computers have transformed our world over just the last few decades and there is no reason to assume the trend will not continue. Computing power is the backbone of modern science, storing, organising and analysing far more data than people possibly could by themselves. In our everyday life too there is now a computer on virtually every desk in every office in the Western world and in almost every home. And, of course, computers power the Internet. These are incredible changes to the way we live and work in the space of just a few years.

After the Second World War computers doubled in their capacity every two years. Then during the 1980s this doubling speeded up to every 18 months. From 1990 onwards the doubling has been occurring almost every 12 months. As well as this increase in power, computers have been shrinking in size and cost at virtually the same rate. Computer power is measured in millions of instructions processed in a second (or MIPS). Around 1990 the fastest computers were running at about 10 MIPS. It has been estimated that the equivalent processing power of the human brain is about 100 million MIPS. By 1998 we had large multiprocessor supercomputers running at a few million MIPS. These, however, cost around 10 million dollars or more and could only be used for big projects like weather forecasting or other scientific modelling.

Through the years there have been many warnings about reaching the end of the possible miniaturisation of computer circuits but each time the stated problems have been overcome by new innovations and the rate of power growth, computer size and cost reduction continues today unabated. It will not be long before that 100 million MIPS mark becomes available not just on large expensive machines but on small affordable ones.

Processing power is one thing but how one uses it is another matter. Have programmers been able to keep up to date with the advances in hardware? One area where there have been huge advances is in programming computers to play chess. Back in 1988, Gary Kasparov (the then world chess champion) was asked if a computer would be able to beat a chess grandmaster by the year 2000. 'No way,' he replied. Later that same year Deep Thought (a specialised chess-playing computer) beat grandmaster Brent Larson in a tournament in America! Then in 1996 Kasparov himself had a match against Deep Blue, which was a scaled-up version of Deep Thought. Kasparov lost the first game of the series but then learnt how to play against it and drew two and won three to finish the winner by 4-2. The very next year, however, he played an improved version of Deep Blue and this time he lost the match 3.5-2.5, a milestone for machine-kind!

Deep Blue used purpose-designed chips running on a mini supercomputer. These days every ordinary PC can run (surprisingly small) chess-playing programs, which can easily beat the average club player and would give a grandmaster a good game. It is interesting to consider how these programs work and how that differs from the way in which we think. There is the story of an Australian grandmaster who, when told that a particular chess program looked at so many millions of positions a second, replied, 'I only look at one position a second, but it is the right one.' He was pointing out that computers do most of their work by brute force; they have to look at every possible move and its consequences even if it is a move that a human player could see at once is no good.

One of the things we are talking about here is the difference between strategy and tactics. Computers are brilliant at tactics, that is short-term thinking considering all possibilities along the lines of 'if I do this and he does that and then I do this' and so on, then evaluating if either side would come out with a better position (e.g. material up). They are much worse at strategy, which is about long-term aims. Things like getting a pawn to the other end to turn it into a queen. It will often be that the actual queening is say 20 or 25 moves ahead, far too distant for the computer to read all the possible intervening moves, but a human will move a pawn forward anyway, if it is safe to do so, because she is working towards that strategic goal. She will also be aiming to move other pieces to support that pawn advance.

Programmers can do some of this by including rules like 'if no other moves bring material gain then push a pawn forward as long as that doesn't incur a loss'. But humans are far better at this kind of strategy. For example, there is an (artificially created) chess position where black has a lot of material advantage but he can't get his pieces to the white king because the pawns are all gridlocked across the board and he can't do anything to break that gridlock. Obviously a draw, the white king can just keep moving around behind his pawns and black can't get to it. White could take one of black's pieces but that would break the gridlock and allow black to win. Any human player would just look at the position and agree a draw. All current computer chess programs, however, as white, would go for the quick gain of taking the black piece even though it breaks the gridlock and allows black to win. The end result of black winning was just too far ahead for the computers to read.

The best of today's chess playing computers look ahead about 14 moves and it is the raw power of modern computers that allows this. This means looking at millions and millions of positions. Programmers have been able to reduce this by what they call pruning. They program the computer to only look a little ahead if it becomes clear there would be a big loss that way. Then they only have to look at the more promising lines to greater depth. Now that computers look so far ahead, however, it has shown up how many tactical mistakes grandmasters actually make when playing in tournaments under time pressure.

When Kasparov or other champions play against computers, they are aware of these limitations and specifically try to exploit them. Computers for instance are suckers for a sacrifice, they think they are getting some material advantage, where the human can see a strategic advantage from the position, possibly to win back greater material in the future. But that sacrifice must be well worked out, if there are any tactical flaws in the human's reasoning the computer will be ruthless.

When playing a computer at chess or some other game, one often drifts into the habit of referring to it as 'he'. It is easy to imagine some intelligence or indeed some motivation or intention in your opponent. Even Kasparov, in that 1997 match, complained that the programmers must have been altering Deep Blue's program while the match was in progress, so convinced was he that there was 'intelligence' behind some of the moves. Of course, this was not the case but then how would we recognise such an intelligence if it was there?

The great British mathematician, code breaker and computer scientist Alan M Turing, one of the fathers of modern computing, proposed a test for computer intelligence that has since become known as the Turing test. The idea is that there is a computer in one room and a person in the next room and they can only communicate via a keyboard and a screen. The person can ask the computer anything and can have a conversation with it via written text. If the person cannot tell whether there is a computer or another person at the other end then the computer has passed the test, it must be considered intelligent. You might think that it would be easy to tell the difference but we are quickly getting near the time when it will be very difficult.

There was a case recently where a psychiatrist in America, who was also a good computer programmer, wrote a dialog program that used many of the techniques and phraseology that psychiatrists usually use. It would start by asking something like 'How are you today' and then, every time you type an answer, it would pick out relevant words from what you said and make up a new sentence using your words and asking how you feel about what you said. Occasionally it would prompt you or say things like 'Go on' or 'Can you expand on that?'. He decided to try it out on his secretary and very soon she was deep in conversation with it. Then she turned to him and asked him to please leave the room because she wanted to discuss personal things with it!

There are now many such dialog programs in existence, built with a knowledge of grammar, syntax and dictionary definitions and including a database of information on how to converse and what type of questions go together. Some of these have even been let loose in Internet chat rooms and it can be quite difficult sometimes to know if you are conversing with a real person or not.

One of the main aspects of these types of interactive systems is the database of knowledge that the program has access to, as in the knowledge of grammar and syntax in the chat programs above or a library of openings and endgames in the case of the chess programs. These 'knowledge based' or 'expert' systems are becoming more prevalent as more and more information is being stored on-line. A good example here is where doctors use expert diagnosis programs to help their own identification of a patient's illness. The doctor will type in the patient's symptoms and medical history and the program will consult its database and come up with a number of possible diagnoses, suggestions for further tests and recommendations for medication and treatment. These mostly agree with the doctor's own diagnosis and will give him more confidence but they do point out things that the doctor might have overlooked. As these get better they might well be used in places where there is not a fully qualified doctor around, for instance in a third world country or maybe even on a mission to Mars or somewhere.

Another development giving a recent string of successes has been in the field of mathematics. There have long been a number of mathematical theorems that mathematicians have believed to be true but could not prove, like the 'four colour' theorem which states it should be possible to colour a map using only four colours to shade the different countries without getting two countries of the same colour next to each other. This was finally proved in 1976 by Appel and Hakin using a computer to do a lot of the work. This was the first computer proof of a theorem that could not be directly verified by other mathematicians, since then there have been quite a few more. There have also been a number of attempts, with some success, to recreate smaller proofs from first principles using machine reasoning or AI (artificial intelligence). The problem with larger proofs is can we trust them? But then that has always been a problem with mathematical proofs that run to many hundreds of pages of logic, how accurate is human verification? Others complain that the computers are using brute force and that the mathematics is not elegant (a bit like the chess computers) but in maths all the stages probably do have to be specified in detail, computer or not.

A particular problem for expert systems is that they are often expert in only one thing, a chess program for instance could not perform a mathematical proof or diagnose an illness, still less make a cup of tea! They lack what we would call common sense, a store of general knowledge about the world and the way it works. We know instinctively for instance (well probably just from long experience) that if you throw something up it will pretty soon come down again. Some AI groups around the world are trying to build databases of this kind of general knowledge for programs to use, like they use a database of chess openings. Other groups are creating programs which learn things for themselves, deduce their own rules and then test them to see if they hold. Some of these programs use 'fuzzy logic', that is making decisions (or best guesses) based on limited data, somewhat similar to the way humans think.

As we have seen in chapter 10, a very large part of our own brain is given over to processing the senses, especially vision. Again, a lot of AI groups are working on vision for robots. This is fundamental to a robot being able to interact with the world, to navigate through it and to gain data about its environment. Good work has been done on pattern recognition using neural networks and, after training, an AI visual system can often pick out a face it knows from a selection of faces, even if the picture is taken from a different angle.

To move around without colliding with things the vision system needs to work rather like ours. It needs to recognise edges of objects, depth perception, slopes, textures etc and to match these things with its database of stored or learnt knowledge about the world. This parallel processing and distributed programming needs to be combined and brought together with fuzzy logic decision making software in order for the robot to interact with the world in real time, as any animal does with ease.

Initially, most robots will be built for single purposes, like the ones that build cars on a production line. But as computing power increases and size reduces, robots will become more general purpose machines and they will be able to (and have to) take decisions for themselves. As I write this, there are two robot explorers, called Spirit and Opportunity, moving around on Mars. They are brilliant machines and we have learnt a huge amount about Mars from them. They have to be told what to do each day but they do have a lot of autonomy too, they must drive in the direction told but they can decide to move round obstacles themselves or stop and wait for more instructions if they sense danger like a crevice in front of them.

Of course, such robots are not yet in our league. There is a story of a researcher not long ago testing a robot rover destined for Mars.

One of its main jobs was to look for signs of life and the researcher was testing it out in a desert in America. As he watched, the robot went about its business, digging a scoop of soil and dropping it into an onboard laboratory for numerous different chemical tests to be performed on it, yet the researcher could see not far off a weed growing in the soil! Obviously the robot could make nothing of that, it wasn't part of its programming.

An interesting question is, what would it take for that robot to have recognised the weed as 'life' when it had not been programmed to know about weeds? Obviously a lot more computing power but that is coming (and soon). A lot more AI capability like vision systems and decision-making, this is being worked on. But would that be enough? Would it need something more – consciousness for example? Well, is an insect conscious? A mouse? How about a dog? It seems to me that consciousness, like free will discussed earlier, is probably a spectrum of awareness. One could argue that it doesn't matter if we call a robot conscious or not, if it passes the Turing test then we just have to act as if it is. This is a bit of a cop-out but at least from a practical viewpoint it would probably work. I want to suggest, however, that there is one thing that animals have that robots don't and that is drives or instincts. We need to do things, like find food, shelter and a mate. We are curious about our environment and hence we explore it, we get hungry and have to eat and we are driven by our emotions. I strongly suspect that these drives and emotions are a large part of what gives us and other animals 'consciousness', however we come to define it.

So, could a robot be given such drives? The answer has to be yes but that this could be dangerous. A robot needs power like we need food. In the house, for example, it could just go and plug itself in if its battery was low (if it felt low?). We have a reward centre in the brain which is stimulated when we do things like eating or making love. Could we build some such subsystem into a robot brain, reward it for being curious, for exploring or, more importantly perhaps, for helping people or following orders? Dogs certainly like being praised, that's how we train them.

The dangers here were recognised by the great science fiction writer Isaac Asimov. If robots become intelligent enough to make decisions for themselves, what would stop them from hurting people, even by accident? He came up with three laws of robotics, which would be hard-coded into the brain of every robot made so that they would have to follow them.

1. A robot may not injure a human being, or, through inaction allow a human being to come to harm.
2. A robot must obey the orders given it by human beings except where such orders would conflict with the first law.
3. A robot must protect its own existence as long as such protection does not conflict with the first or second law.

Science fiction, yes, but we must think about some of the consequences of intelligent, autonomous, non-humans at large in society. What would be their status? What if a robot caused a fatal accident? Who would be blamed, the robot, his 'owner', the manufacturer?

I touched on robotic life in chapter 8 on ET. Statistically it is highly likely that other civilisations have arisen in our galaxy prior to ours. That chapter pointed out, however, the immense difficulties with organic life spreading to other star systems, which, due to inevitable natural disasters, is the only way for intelligent life to survive in the long run. Machine based intelligence would not have the difficulties associated with biology and it seems reasonable to assume that it could reproduce, improve itself through research and spread through the galaxy.

Should we be worried? It does seem likely that our machines will survive us and that this has probably already happened elsewhere in the galaxy. What motivations will they have? How will they treat us? We could of course shape these machine motivations in the first instance but in the longer term they may well redesign themselves in many different ways. If we remember back to chapter 10 we met game theory and this showed that, logically, it is a better result all round to co-operate with others than to oppose them, machine intelligence should be able to appreciate that. But further, it is my opinion (wishful perhaps?) that such an intelligence, created initially by us or by aliens, would recognise and value organic based intelligence as kindred spirits. In a sense, robots could be more human (humane) than we are ourselves.

In conclusion for this chapter, while humans are not currently becoming more intelligent, computers definitely are. Projecting forward from the current trends given at the start of this section on artificial intelligence and considering advances in AI, we will have cheap robots with human level intelligence by the year 2030. That's less than 25 years away! I wonder what they will say to us when they communicate? Indeed what will they think of us with our poverty, wars and exploitation of others? There is an old Chinese curse which says 'may you live in interesting times'. It is certainly going to be that!

Chapter 13

Conclusion

The mist of familiarity obscures from us the wonder of our being.

Percy Bysshe Shelley

Atheism leaves to man reason, philosophy, natural piety, laws, reputation, and everything that can serve to conduct him to virtue; but superstition destroys all these, and erects itself into a tyranny over the understandings of men.

Francis Bacon
(English politician and philosopher 1561 -1626)

Man is a credulous animal, and must believe something; in the absence of good grounds for belief, he will be satisfied with bad ones.

What is wanted is not the will to believe, but the will to find out, which is the exact opposite.

Bertrand Russell

The previous 12 chapters of this book might seem an eclectic collection, delving as they do into mythology, space, time, physics, biology, extraterrestrials, culture, evolution, psychology, the scientific method and even an attempt to predict the future. The underlying theme, however, has been that science, in each of these fields, has in just the past 100 years or so uncovered realms that were undreamt of before. Counter-intuitive ideas that dive beneath *the mist of familiarity* and bring back *the wonder of our being*.

As mentioned before, the twin themes of this book are 'Where did we come from?' and 'What is the nature of reality behind our perception of the world?'. The twin messages I would like you to take away are 'evolution is not about progress' and in Shelly's words 'the necessity of atheism'.

But don't just take my word for it – read; read as much as you can about these fascinating ideas and do so from many different sources. And then make up your own mind. Don't take what one person tells you as gospel; ask what their evidence is and if they say such and such can't be questioned, be rightly suspicious!

I have tried in these pages not only to present scientific information but also in many instances to address the question 'How do they know that?' Frequently these ideas are based on long chains of reasoning but more often than not they are also arrived at from more than one direction. The more independent lines of argument there are supporting a theory the more credence can be put into it, for instance see chapter 4 on the Big Bang or chapter 6 on evolution, both of which over the past 100-200 years have moved from initial speculation to mainstream scientific fact. More is being learnt all the time of course but this is just filling in the details, the broad picture of both of these is now as accurate as we can paint it.

Science, as pointed out in chapter 11, is a collaborative enterprise. It is amazing what can be accomplished by people working together, checking each other's results and building on previous research. Just think for instance about the chess program Deep Blue which beat the world champion. First teams of people creating the computer and continually improving it and then a team of people creating the chess-playing program which finally beat Kasparov. People working together can do far more than our best talented individual working alone.

It has become fashionable recently to label atheism and indeed science itself as just another belief system, one way of looking at the world out of many. This is not true. Science is about the absence of belief, it is a world view that is free from superstition and mysticism. Scientific ideas are checked, verified and are updateable, that is the nature of the enterprise.

Listen to the following quote from physicist Bob Park: –

> 'Pattern recognition is the basis of all aesthetic enjoyment, whether it is music, poetry or physics... Unfortunately, the brain that makes the link between the tides and the phases of the moon may also connect a comet to victory in battle. Science is about spotting real patterns.
>
> 'Richard Feynman described science as "what we have learned about how not to fool ourselves". Science depends on openness: we expose our findings, including the details on how they were obtained, to the scrutiny of the scientific community. This sounds like a prescription for chaos, but the result is the opposite because it reinforces the idea that science is conditional – always subject to being replaced by better information. This can be frustrating to non-scientists, who ask why science can't make up its mind, but the alternative is dogma. Openness provides a mechanism for self-correction, setting science apart from other ways of knowing. Science is, in fact, the only way of knowing. Anything else is just religion, which is all about authority.'

Bob Park

To return to one of this book's main themes, the nature of reality behind our perception of the world. Take the following fact that science, with its modern tools, reveals to us. Every second, 100 billion neutrinos, originating from the sun, pass through every square centimetre of your skin and not one of them is likely to be stopped by the atoms in your body or your clothes. This is a fascinating

162

example of how empty apparently solid mater really is and how little we actually see of what's around us.

Evolution has shaped our perceptions of the world to be that which is most useful to us for surviving and reproducing. Solidness is just how we interpret force fields, matter is mostly nothing! Colour is only particular wavelengths of electromagnetic radiation, there is no such thing as 'red', it is just our perception of a particular wavelength of photons translated into chemical and electrical signals in our optic nerves. Also, there are many wavelengths of light in the world that we can't see.

From chapter 10 on evolutionary psychology we learnt that not only is our perception of the world shaped by our evolutionary history but that so too is our behaviour, our moral judgments and the very way our brains think about things. We also learnt that the very strong feeling we have that there is an 'I' inside us, viewing the world and making rational decisions about things is really just an illusion, it is a story we tell ourselves after the fact. There is no mind which is separate from matter, there is an incredibly complex brain, bathed in a cocktail of chemicals, indelibly interlinked with the body and functioning together with it as a single unit.

You could get on with your life without knowing any of these astonishing ideas. In fact that's exactly what we're equipped to do, but the information is there and we, uniquely, have the capacity to understand it even though much is counter to our normal expectations about the world. As Douglas Adams says, 'We live at the bottom of a deep gravity well on a ball of rock rolling through space going round a huge nuclear furnace 93 million miles away and we think this is normal!'

Another concept we met, in chapter 3, was the incredible length of deep time, the idea that our own three score years and ten, indeed all our history as a species, is no more in cosmic terms than a passing split second, an afterthought in earth's long history. Unimaginable eons passed before multicelled animals appeared, let alone our short dance upon the stage. It is bacteria that rule the earth, they have been here for millennia prior to us and will be here long, long after we have departed and all our achievements have crumbled to dust.

This leads us on to our second theme of 'Where did we come from?' And the reason it does so is due to deep time. Before 1900, although evolution was pretty much accepted there was the problem of there not really being enough time available for it to occur. After nuclear physics was worked out, however, it became clear that the earth was actually billions of years old and that there was time aplenty for natural selection to work its magic. This is another example of the collaborative and progressive enterprise of science, with its different fields and disciplines overlapping and supporting each other.

If you consider a human being in any depth at all you will be rightly awed by our complexity: an incredibly dextrous, tool making, intelligent and caring animal, capable of putting a telescope in orbit and writing a Shakespeare play or a Mozart symphony. It is an entirely understandable reaction to say something like, 'Look at how complex we are and how well designed, I can't believe that we arose by blind chance.' (Dawkins calls this the argument from personal incredulity). Of course, if you put us side by side with a rock there is an incredible difference. Even if you put us side by side with another animal it is hard to imagine the leap from one form to another. The problem here is that over the years the intervening forms have disappeared. There is a big gap, for instance, between us and the chimpanzees (our closest living cousins) but there would be a much smaller gap between us and say Homo erectus if they were still here. Much more believable is that it needed just a few minor changes in a few of our genes to take us from cave men to modern people.

Now we reach the main point of this book and the reason for the wide diversity of subjects collected here together. Starting with the complexity of Homo sapiens is the wrong way to look at things, there is just a small step from us back to Homo erectus. There is an equally small step back to the immediate ancestor of Homo erectus. In fact there are comparable, believable, small steps all the way back to the beginning of life on earth. And even then, all the heavy elements that make up the rocky earth and the chemistry of life on it came from the nuclear powerhouses of supernovas and before that the hydrogen and helium for these first stars was generated in the Big Bang. Small believable steps all the way. We may not yet fully understand every last chemical interaction in every step but the main outlines and indeed an incredible amount of the detail is understood and documented widely for anyone to read. So looking at man and saying 'unbelievable' is the wrong way round, it is the Big Bang that seems to have no cause; the rest, if not predictable, is at least logical.

Now you could say, 'I can't believe the universe came from nothing with the Big Bang.' Well, we have no information about that (although we are working on it). But, given the Big Bang, all else follows in easy steps. That is not to say that we are inevitable, because of the large part chance plays (as Stephen Jay Gould says 'If you replayed the tape of life it would not repeat itself and we would not be here, although it is entirely possible some other type of animal would achieve intelligence') but there is definitely no need to invoke a 'guiding hand' of any sort. All the steps from the Big Bang to the present are logical consequences, as demonstrated throughout this book, indeed to postulate a guiding hand giving nudges here and there in some preferred direction would in itself require further explanation and would undermine our confidence in the way nature works.

Some suggest an alternative, that God set the universe in motion by creating the Big Bang and all the laws of nature but then did not interfere after that. To what purpose? To wait for us, the pinnacle of creation? He must have sat about for around 13.5 billion years waiting! This again is the conceit to see humans as special or chosen, the Victorian concept of progress in evolution, instead of just another animal with perhaps a certain deftness in tool use and a slightly larger brain. It is reverting to magic to cover an area that we don't know about.

> *God is an hypothesis and, as such, stands in need of proof: the onus probandi [the burden of proof] rests on the theist.*
>
> **Percy Bysshe Shelley**

Some people find this view of life depressing or empty, no purpose, no guiding hand looking after you, no God given values of right and wrong. I strongly disagree, I find it much more comforting that the laws of physics are not something that can be cast aside whenever some supernatural entity decides to intervene. And it's the opposite of depressing, indeed a scientific outlook is often based on a deep awe and wonder at the universe.

But what about values? If there is no intrinsic right and wrong, how will we know how to behave? What will stop us from exploiting others? The people who usually say this point to the Bible (or other holy text) as the guide to moral living. Who are they kidding? The Bible, especially the Old Testament, is littered with violence to others, subjugation of women and other barbaric acts. When particular horrific passages are pointed out to the speaker they may well say that it was different in those days or that the story is just a metaphor or that the New Testament is more relevant. In other words, they are not getting all their morals from the text but they are picking and choosing the ones that seem to them to be worthy. How are they making that choice?

It can only be coming from within them. Indeed, it is in us all. Just think for a moment. You know what's right and wrong. Is it wrong to kill someone for no reason? Is it wrong to take advantage of people? Of course, it's wrong and you don't need someone else to tell you that, you just know it, even a criminal behind bars knows what's right and wrong if you ask him. So where does this inbuilt sense of right and wrong come from? It may well be that we could postulate some logical, intrinsic reason why it is wrong to kill, one that any intelligent being (including say an intelligent computer) would agree with. However, if we remember back to chapter 10 on evolutionary psychology, we learnt that much of our behaviour comes from the fact that we are a gregarious, group-living species who forms strong, mutually dependant and mutually beneficial bonds with other group members. We keep track of who owes what to whom and we have a strong sense of fair play in our dealings with each other. Also we have those mirror neurons which give us empathy with others when, say, they are suffering. Much of this is also reinforced by our family-based upbringing as we learn how to interact with others and develop self-control.

On the occasion of disputes, we have created a system of law that helps to protect the weak and vulnerable from being exploited by people who are maybe lacking some of the areas of empathy for others, either due to genetic deficiencies or an unbalanced and impoverished upbringing.

You know what's right and wrong as much as any priest in any pulpit and you don't need someone setting themselves up saying they have divine guidance and telling you what to do. Society, though, does have rules and laws for people living together in close contact. These are a consensus that has built up over the years and, although long-winded, they are probably very wise. Having said that, the group consensus does change subtly across the years, responding to developments in the structure and size of society; for instance we now don't have the death penalty in the UK and we probably now view the time when it was in place as a slightly more barbaric one.

Where does all this leave us? Well, we are moral animals in a lonely but fascinating universe. We are just at the beginning of a fantastic journey of discovery, if only we can solve the problems we are creating for ourselves and protect the planet and its diversity of life for future generations. Probably the best thing that could happen to us would be the discovery of life outside our solar system. That would a) confirm that we are not special and b) might bring us more together as a single humanity.

So there is no reason for us being here, apart from the fact that all of our ancestors were good at surviving, together with their blind luck at avoiding accidents and catastrophes, an unbroken line stretching back to the dawn of life. We have no purpose in our short insignificant span here on earth other than that which we choose to give to ourselves. Science and its tools have probed beneath our illusions of the world and have given us glimpses of an alien reality, often in direct opposition to our common sense experience. And we now know that we are literally Stardust, dust from the stars, endowed via natural selection with the gift of intelligence, with which we can study this remarkable universe we find ourselves in.

Next time you look at a sunset, take a moment to remember that the sun (our local star) is not really sinking to the west, that is an illusion created by the world spinning west to east on its axis. So, revel in the colours of a sky ablaze, of the clouds on fire, but also, spread your arms and open your imagination, experience the whole world and you on it rolling backwards in space as it cycles on in its endless journey round the sun. Understanding can only add to beauty.

Chapter 14

Recommended Reading

Reading is to the mind what exercise is to the body.

Sir Richard Steele (1672 – 1729)

I have only scratched the surface of the subjects covered in this book. Each chapter could have been a full book in its own right. What I have tried to do is pull together an overview of the most recent scientific thinking relating to what we are, where we came from and the nature of the reality of the universe beyond that which we perceive with our naked senses. I hope I have stimulated the imagination and wetted the appetite to find out more. In that respect, I can thoroughly recommend the following authors and their books for more in-depth discussions on some of the themes touched upon in this book.

Anything by <u>Richard Dawkins</u>: All his books are entertaining and instructive. A good introduction to his work would be ***River Out of Eden***. But ***The Selfish Gene*** 30th anniversary edition, ***Unweaving the Rainbow*** and ***Climbing Mount Improbable*** are also highly recommended.

Anything by <u>Stephen Jay Gould</u>: All his books are a delight to read, no one is his equal in terms of breadth of knowledge and use of language. Many of his books are in the form of a collection of his essays, of which he is a master. A good introduction might be *A Dinosaur in a Haystack*.

Note:
While Gould and Dawkins are both atheists and strong Darwinians they have in the past had their, well publicised, differences. Gould is more tolerant of religion than Dawkins; he thought it and science were two non-overlapping fields of study. Dawkins does not. Gould also found it difficult to accept natural selection acting at the level of DNA rather than at the whole organism level. In both these differences I side strongly with Dawkins but having said that it is extremely instructive to get both points of view. Lively discussion of differing ideas is the lifeblood of scientific progress. Do read Gould, he's brilliant.

Daniel Dennett: A philosopher with a number of books to his name but I would especially recommend *Freedom Evolves* as a good introduction to his thinking.

Stephen Pinker: Again he has a number of books, all good, but I strongly recommend *How the Mind Works* and *The Language Instinct*.

Matt Ridley: A science journalist with a number of fascinating books, especially *The Origins of Virtue* and *The Red Queen*.

Steven Johnson: Again a science journalist. I recommend *Emergence* and *Mind Wide Open*.

New Scientist: A weekly magazine covering developments in science and technology with a strong social conscience. Well worth subscribing to.

Douglas Adams: He is better known for his hilarious science fiction but, after his death, a collection of his essays and other writings

were published in a book called *The Salmon of Doubt*. No
should be without it.

Oliver Sacks: A neurologist studying how illness and injury in
brain affects behaviour and perception. See *The Man Who Mistoo*
his Wife for a Hat Stand.

Richard Forty is a senior palaeontologist at the Natural History
Museum in London. His writing is both informative and entertaining,
especially *The Earth: An Intimate History* and *Life: An*
Unauthorised Biography.

Jostein Gaarder: A philosophy teacher from Norway. His book
Sophie's World is a fascinating mixture of mystery story and history
of philosophy, written for anyone from 15 or 16 upwards.

Edward O Wilson: One of America's foremost scientists, a biologist
and Harvard professor who has written many fascinating books
including *On Human Nature*, *The Ants* and *The Future of Life*.

John Gribbin: A British astrophysicist who has written many books,
one of which, *The Birth of Time*, is a fascinating account of how
we came to measure the age of the universe.

Stephen Hawking: One of the most brilliant theoretical physicists
since Einstein. His *A Brief History of Time* is a classic that should
not be missed.

Sam Harris: An American philosopher specialising in neuroscience.
His book *The End of Faith* was a best seller in the States, after
which he received hundreds of letters from Christians who disagreed
with him (often in the most hostile terms). His new book *Letter to a*
Christian Nation is his response to them in the form of a letter
back. This little book superbly demonstrates that people don't get
their morality from religious texts like the Bible and that where they

171

...ften a very harmful morality that they end up

...ion, I could go on and on. Too many books ...t in the available treasures but surely a ...han too few! I can't finish, however, without ...ding one more author – <u>Shelley,</u> my favourite poet. Some ...as just soar and soar and only poetry can express them, so temper your science with words from the arts and fully explore the *shapes that haunt thought's wildernesses.*